AF880718

Royal Bavarian Jagdstaffel 76

Its History
Its Pilots
Its Aircraft Markings and Colors

Bruno Schmäling

Reinhard Kastner & Jörn Leckscheid

Royal Bavarian Jagdstaffel 76

Its History
Its Pilots
Its Aircraft Markings and Colors

Bruno Schmäling

Reinhard Kastner & Jörn Leckscheid

This book is dedicated to Dr. Walter Böning and all the members of Royal Bavarian Jagdstaffel 76.

Translated by Catherine Lawler & Jörn Leckscheid

Design and layout: Jack Herris
Cover design: Aaron Weaver
Front cover painting: Russell Smith
Color profiles: Bob Pearson
Digital photo editing: Aaron Weaver & Jack Herris

Aeronaut Books

Books for Enthusiasts by Enthusiasts

www.aeronautbooks.com

Publisher's Cataloging-in-Publication data

Schmäling, Bruno
Royal Bavarian Jagdstaffel 76 / by Bruno Schmäling, Reinhard Kastner, and Jörn Leckscheid
p. cm.
ISBN 978-1-953201-43-0
1. World War, 1914–1918 -- Aerial operations, German. 2. Fighter pilots -- Germany. 3. Aeronautics, Military -- Germany -- History. II. Title.

ND237 .S6322 2011
759.13 --dc22 2011904920

Above: Walter Böning with his dog in 1918.

The Authors

Bruno Schmäling has dealt intensively with the history of the German air service in the 1st World War for many years. In the 1970s he interviewed 70 former members of the German Jagdstaffelen of the 1st World War. He is the primary author of *Royal Prussian Jagdstaffel 30, Royal Bavarian Jagdstaffel 23,* and *Jasta Colors Volume 1*.

Reinhard Kastner started to research the history of Bavarian military aviation 1912–1918 decades ago and owns the biggest private archive on Royal Bavarian military aviation. He wrote a large number of articles for *Over the Front* and he works as editor of the *Propellerblatt* in Germany.

Jörn Leckscheid has had a life-long interest in aviation. His interest eventually bacame a serious hobby, which lead to a long cooperation with the late aviation historian Alex Imrie. He is the co-author of *Jasta Colors Volume 1* and will be the co-author for further future projects.

Collaborators

A book such as this is never the result of the authors' efforts alone, but rather always represents the teamwork of a number of individuals. For this reason, our thanks are due to the generous and dedicated collaboration of the following:

Rainer Absmeier
Winfried Bock
Lance Bronnenkant PhD
Jon Guttman
The late Alex Imrie
Andrew King (Old Rhinebeck Aerodrome
Dr. Volker Koos
David Méchin
Colin Owers
Tom Polapink (Old Rhinebeck Aerodrome)
The late Les Rogers
Greg VanWyngarden
Aaron Weaver
Tobias Weber
Reinhard Zankl

A very special thanks for great cooperation goes to Philippe Charroin and the staff at the "Aéroplane Restauration Management" at Le Bourget, France.

Acknowledgements

A debt of gratitude is owed to those former pilots of the 1st World War who shared their knowledge and experiences through interviews and correspondence. These include Herr Dr. Walter Böning (Jagdstaffels 19 and 76) and Herr Otto Fuchs (Jagdstaffels 30, 35, and 77), and Herr Heinz Arntzen (Jagdstaffels 15 and 50).

A number of individuals closely related to the personalities who are the subject of this book also deserve our grateful acknowledgment of their contributions. These include the late Frau Anna Vollert, widow of Clemens Vollert, Herrn Polizeipräsident a.D. Koller, son of Vzfw. Karl Koller, the late Frau Agnes Schwörer-Böhning, widow of Hans Böhning and her sister-in-law, Frau Ruth Schwörer, the late Frau Wenig, widow of Erwin Wenig, the late Frau Wildmann, daughter of Josef Bauer, as well as the relatives of Julius Geyer and Karl Hopf.

A very special thanks are due to Frau Dr. Haggenmüller, director of Department 4 of the Bavarian Main State Archive in Munich, as well as her co-workers, for their dedicated support during the research for this book.

The authors owe a big thanks to cover artist Russell Smith, who brought to life the special atmosphere of a morning patrol along the Vosges Mountain range by a trio of aircraft from Jagdstaffel 76 in his cover painting. Our special thanks go to Bob Pearson for all the color profiles for our book, to Mrs. Catherine Lawler for the translation, and, of course, a very special thanks to our friend and publisher Jack Herris.

Bruno Schmäling, Reinhard Kastner, Jörn Leckscheid.

Last but not least I want to extend my personal heartful thanks to my wife Kerstin for all her support and patience while I collected the material over decades and worked on the World War I aviation books.

Bruno Schmäling

Table of Contents

Foreword

The following book is the third volume in a series covering the history of the German Jagdstaffeln of World War I.

The nucleus of this book was provided by the letters written by Staffelführer Walter Böning to his parents and relatives during his time of service at Jagdstaffel 76. These letters provided a highly detailed insight into the frontline service of the Staffel. Furthermore, they also vividly illustrate the difficulties and obstacles a Staffelführer of the so-called "America Program" had to face in 1918. The letters were written immediately following the events and therefore provide unique and first-hand historical insight. The information provided by the letters was supplemented by the interviews Bruno Schmäling conducted with Dr. Walter Böning in September and November 1976.

A very interesting aspect of the history of Jagdstaffel 76 was provided by the "Aéroplane Restauration Management", at Le Bourget airfield in France. Fortunately, just four years prior to the writing of this book, the remains of an Albatros D Va operated by Jagdstaffel 76 were discovered in a swamp near the village of Boussicourt.

As with all of our books, this one is also based primarily on extracts from Jagdstaffel War Diaries transcribed by Herr Erich Tornuss of Weissenfels in the 1930s. In spite of the gaps left in these transcriptions, they are of immeasurable value as the original Jagdstaffel war diaries, with two known exceptions, were destroyed in the archive at Potsdam by Allied bombs during the closing stages of World War II. The "Tornuß-War-Diaries" were carefully reworked and supplemented by Dr. Gustav Bock, and Winfried Bock made these available to aid a number of projects by the authors. Additional information was provided by the relatives of several former Jagdstaffel 76 members, who generously supported the research with documents and photos.

Another very important foundation for the book are the documents found in the Bavarian Main State Archive, Department 4 in Munich, especially the personal files of the officers, the Kriegsstamm-Listen (parent units lists) and the Kriegsstamm-Rollen (parent unit scrolls) of the members of Jagdstaffel 76.

New light on the final two months of the war at Jagdstaffel 76 was shed by a number of highly interesting photos provided by Greg VanWyngarden, Rainer Absmeier, and Tobias Weber who generously made them available for the book.

The maps used for the book are taken from original German World War I military maps. As these maps mainly cover the frontline area, in some rare cases airfield or locations of victories or losses cannot be pointed out on these maps.

After the passage of more than a century, and the loss of a large number of original documents in the Second World War, there are of course gaps and open questions, which is almost inevitable when taking on such a project. Therefore, the authors would be grateful for any sort of supplemental material or additional information that may provide more insight into the history of Royal Bavarian Jagdstaffel 76.

Bruno Schmäling, Reinhard Kastner, Jörn Leckscheid.
Winter 2021

Prolog: Jagdstaffel 19

St. Loupe en Champagne, 7th German Army

On October 2, 1917, the Jagdstaffel 19 Albatros D V slowly floated toward the grass strip of the Saint Loup airfield in the Champagne region. One by one they touched down and taxied toward the sheds where mechanics waited to refuel, re-arm and conduct inspections on the aircraft, searching for possible damage by enemy bullets.

One of the pilots, Lt. Walter Böning, taxied his Albatros D V towards the mechanics. It was marked with a blue and white fuselage band and the black letter "B" painted onto the fuselage sides. He unfastened his seatbelts, climbed out and turned to the mechanics to briefly explain:

"We sighted several Spad and Caudrons; they turned away, however. I sent a few shots after them. Everything should be alright otherwise."

He then joined the other pilots on their way to have lunch at the Officer's Mess. Soon after he had taken a seat, an orderly entered the Mess and heading directly toward him:

"Lt. Böning, I just received a telegram; you have been appointed to become the Staffelführer of Jagdstaffel 76."

Walter Böning looked up in complete surprise. Of course, he was aware of the America-program and the intended formation of several additional Jagdstaffeln as a consequence of it. But he had not expected to be named Staffelführer of one of these new units. He read the telegram, which contained orders for immediate transfer to Schleißheim in Upper Bavaria, where he was to oversee the establishment of Royal Bavarian Jagdstaffel 76.

His current Staffelführer Lt. Ernst Hess was an exceptional fighter pilot who had gained the respect of his flyers within a few short weeks following his arrival as Staffelführer at Jagdstaffel 19. He turned to Walter Böning and said in a humorous way:

"It seems as though we will have to pass you on to your native tribe. As you know, I do not want you to go, but we have to follow the rules of the bureaucrats at higher levels. In any case, take good care of yourself when you join the mountain tribe."

Walter Böning nodded his head. He finished his lunch quietly. Then he took a seat in one of the Casino's comfortable, plush chairs in a quiet corner. Utilizing his sober, analytical and deliberate mind, he was fully aware of the responsibilities he was about to take on. The wellbeing of the Staffel would depend on him as its new Staffelführer, including all of the young, inexperienced flyers, amongst which some untalented ones would be mixed in. He was under no illusion that this task would by far outweigh his previous responsibilities as Jagdstaffel 19 Kettenführer (formation leader of three aircraft).

He looked around the room at his comrades who had also made themselves comfortable in the Casino. Saying goodbye to his Jagdstaffel 19 friends would be difficult. He had been a part of the Staffel since its formation in October of 1916. On April 6, 1917, he had logged the Staffel's first aerial victory. He had made friends, and he knew blindly how to assess each of them on any given mission. Jagdstaffel 19 was a sworn fellowship, thanks in part due to Oblt. Erich Hahn, the second Staffelführer who had been in command of the Staffel since November 1916. He was an experienced pilot and exceptional leader; but above all, he was a fabulous person who taught Böning the importance of every individual's position within when the Jagdstaffel during missions. Erich Hahn practiced Ketten formation flying and Staffel formation attacks countless times, always weighing the prospects of success against its risks. Out of 23 aerial victories (two of which were only acknowledged as "forced to land") credited to Jasta 19 during his leadership, Erich Hahn had only lost one of his men in aerial battle, and this happened while he was on leave. However, Hahn himself was killed in aerial combat on September 4, 1917, falling victim to French ace Geoges Félix Madon, who would survive the war with 41 confirmed victories. His successor, Lt. Ernst Heß right away stepped into Hahn's footsteps, successfully leading the Staffel in the same manner Erich Hahn had done before him.

Walter was determined to lead his Staffel just the same way. He thought about Erich Hahn, who had departed alone about a month ago, and did not return. Böning had visited the crash site and saw the Albatros, its shredded debris scattered in an arc-like pattern. Chills went down his spine at the thought of losing such an amazing person. The sound of the bell pulled him from his thoughts, and the Staffel departed on another flight to the front. Gathering

Above: Lt. Walter Böning in front of his Albatros D V D.2092/17, Jagdstaffel 19 at St. Loup airfield. The aircraft carries his personal markings, a blue-white-blue fuselage band with a black, white bordered letter "B". The nose of the aircraft is painted white followed by a blue stripe; the rudder is black.

Right: Lt. Walter Böning (left), with Oblt. Erich Hahn, the Staffelführer of Jagdstaffel 19, and Lt. Rahn, who left Jagdstaffel 19 on the same day as Walter Böning.

his things and walking alongside his fellow pilots, he proceeded to the airfield, where their aircraft were lined up, fully fueled and resupplied with ammunition.

The following day, he went to the office in order to speak to Royal Prussian Jagdstaffel 36 on the telephone. After having overcome some difficulties, he reached a staff officer there.

"This is Leutnant Walter Böning, Jasta 19 speaking, may I speak to Lt. Hans Böhning?"

The officer on the other end of the line affirmed, and after a short wait, he heard his friend's distinct Bremen dialect. Walter Böning knew Hans Böhning from their time at Erlangen University. They became friends, and Hans Böhning had taken him along to air shows in Fürth, as well as in Munich at the Oberwiesenfeld, and in Schleißheim. He was the one who had awoken Walter Böning's interest in aviation.

"I have been appointed Jagdstaffel 76 Staffelführer. I would be delighted if you joined me

Above: Walter Böning, reading a book at the Jagdstaffel 19 officer´s casino at St. Loupe.

as my technical Offizier. I need someone whom I can fully rely on. Establishing a new Staffel is no small feat."

Hans Böhning realized that his days as a Bavarian officer of the Royal Prussian Jagdstaffel 36 were numbered. Efforts had been underway to remove Bavarian pilots from "non-Bavarian" Jagdstaffeln, transferring them to Bavarian Jagdstaffeln. Saying goodbye to his "36" would be difficult for Hans Böhning. At the same time, the thought of this new assignment appealed to him:

"Alright, you can submit a request for my transfer to your Staffel, I will talk to Bülow (Lt. Walter von Bülow, Staffelführer of Jagdstaffel 36). I would rather join your pack than be transferred wherever else."

Walter Böning was relieved to have gained a comrade for his new Staffel that he could rely upon one hundred per cent. He would now attempt to find one or two more of his old acquaintances to join his Staffel.

Otto Fuchs, fighter pilot serving with Jagdstaffel 30, and Staffelführer of Jagdstaffel 35 and 77 made the following remarks concerning the camaraderie in Jagdstaffeln:

"One got accustomed to the comrades, got to know their personalities and peculiarities of each of them. We spent a lot of time together. Friendships formed, and the trust in our relationships grew deeper. I was fortunate to be met with great camaraderie in all the aviation units I served, making it more difficult to say goodbye."[1]

On one of the following evenings, Staffelführer Ernst Heß held a small farewell gathering for Walter Böning as well as Arthur Rahn, who was also about to leave the Staffel, wishing them luck and success with their new Staffeln before they packed their belongings to leave.

On October 6, 1917, Walter wrote the following lines to his parents:

"Tomorrow I will leave for *Schleißheim. I was appointed Führer of Jagdstaffel 76, but I have yet to establish the Staffel. I could think of better things to do but it is progress, nonetheless."*[2]

On October 7, 1917, an orderly drove Walter Böning to the nearest railway station in the Staffel vehicle, along with Arthur Rahn, who had been transferred to Jagdstaffel 18. Walter boarded the train and embarked on his journey to Schleißheim near Munich.

While talking to Bruno Schmäling about his transfer, he remarked:

"Saying goodbye was hard; had I known the trouble and difficulties that would face me in the following weeks and months, I would have struggled to say goodbye even more."[3]

1 Formation of Royal Bavarian Jagdstaffel 76

Schleißheim, September 1917 – November 1917

Right: Aerial view of Bavarian Flieger-Ersatz-Abteilung I at Schleißheim, north of Munich. In addition to the long row of hangar buildings, three groups of aircraft tents can be made out on the left, in the center and the upper right corner of the photo. (R. Kastner)

On September 7, 1917, Bavarian Flieger-Ersatz-Abteilung 1 in Schleißheim, located north of Munich, received orders for the formation of Jagdstaffel 76 by October 15, 1917. As previously mentioned, Lt. Walter Böning of Jagdstaffel 19 was appointed Staffelführer on October 2.[1]

Walter Böning was born in Oldenburg in the Prussian Grand Duchy of Oldenburg on January 14, 1894. Following his High School graduation, he decided to study medicine, as his father had done before him. He attended the University at Erlangen, a college town north of Nuremberg. According to a remark he made to Bruno Schmäling, he enjoyed the Bavarian lifestyle in spite of being "a Prussian". It was a wonderful, carefree time for him. Located right to the south of Erlangen, Nuremberg's medieval city offered inviting taverns and cozy beer gardens. There were many lakes in the vicinity of Erlangen which made for wonderful swimming destinations in the summer. To the south, the romantic Altmühltal (Altmühl valley) was a great place for hiking. During his time at university, he made the acquaintance of Hans Böhning, another "Prussian" from Bremen, and they became good friends. Hans Böhning was interested in aviation and took him along to air shows which were held at Fürth near Nuremberg as well as Munich's Oberwiesenfeld and Schleißheim. Hans Böhning's enthusiasm for aviation was contagious, and soon, Walter's interest in aviation began to grow, too.

On April 1, 1914, he volunteered for a one-year assignment with the Bavarian 19th Infantry regiment at Erlangen, in order to fulfill his military duty (typically, compulsory military service required 2 years of service). The comfortable way of life in Bavaria took an abrupt ending on August 1, 1914, with the outbreak of the conflict that became the 1st World War. He moved into the field with his Infantry

Left: Lt. Walter Böning as a pilot of Bavarian Feldflieger-Abteilung 6 in front of an Albatros C VII. A little doll resembling his dog has been mounted to the bracing wire over his right shoulder.

regiment. As early as December 21, 1914, he was promoted to Unteroffizier and Offiziersaspirant. On May 11 and October 25, 1915, promotions to Offizier-Stellvertreter and Leutnant der Reserve followed, respectively. The same year, he was awarded the E.K. II, Oldenburgisches Friedrich-August-Kreuz II. Klasse. On May 7, 1915, he received the Bavarian Goldene Tapferkeitsmedaille (Golden Medal for Bravery).

Ever since he had attended the aforementioned air shows alongside Hans Böhning, Walter could not get aviation out of his head.

Down from the trenches, where he found himself soon after the outbreak of hostilities, he kept a watchful eye at the ever-increasing number of aircraft in the sky. He recalled of all the wonderful days when he and Hans Böhning had experienced, as they witnessed the flight demonstrations in Fürth and Schleißheim. It was at that time that thought of becoming an aviator began to manifest in his mind. It was simply a matter of consequence that his logical way of thinking resulted in his decision to apply for transfer to the Fliegertruppe.

On February 10, 1916, he was transferred to the Bavarian Flieger-Ersatz-Abteilung in Schleißheim in order to begin pilot training.

Following a training period that lasted nearly three months, he was sent to Army Aviation Supply Depot A on May 9, 1916. On May 25, he was commanded to join Bavarian Feld-Fliegerabteilung 6 at their airfield in Bühl (French Buhl-Lorraine) near Saarburg (French Saarebourg).

The "6b" comradeship was excellent. Soon, he became friends with one of his observers, Lt. d.

Res. Clemens Vollert; whenever possible, the two flew together on missions as "Emil und Franz" (nicknames for pilot and observer).[2]

It was around the onset of dawn on July 23, 1916, when Walter Böning and his observer, Lt. d. Res. August Ertl were on the way back to their home base following a sortie around Lunéville. Due to an emerging cloud cover and increasingly poor visibility, Böning decided to make an emergency landing in unfamiliar territory; the wings of his aircraft brushed a tree. A few moments later, the aircraft slammed down on a field. Walter Böning was pulled from the wreckage with a broken collarbone as well as fractures to his left ribs, while Leutnant Ertl sustained a concussion and internal injuries. Both were transported to Duss field hospital for initial treatment. On August 6, 1916, they were transferred to the reserve field hospital in Saarburg, which was located nearby the FFA 6b airfield at Bühl. Walter Böning remained there for treatment until August 22, 1916. After being discharged, he was sent home on a three-week sick leave.[3]

On July 27, 1916, Walter Böning wrote a short letter to his family from the field hospital. He reported:

"I am doing well overall. My whole body is aching from being sprained and strained, there are possibly some small fractures. Minor injury to the right eye. I returned very late after a bombing flight against Luneville. We were caught in some clouds, and I had to land. One of the wings hit a tree, causing the aircraft to crash. The observer sustained the same injuries as I did."[4]

In the letter to his father, he withheld the fact that he did not emerge from the crash with the few consequences described, as can be gleaned from another letter:

"I will attempt to write but it is slightly difficult as my left hand is sprained, and my right arm is of little use. As soon as my breathing improves (probably caused by blood in my chest cavity) they plan on treating me with a scalpel before stitching me back up in order to reattach what came loose during the crash, particularly a few tendons on my right shoulder and my collarbone.

The worst part is that I am constantly having difficulty breathing; I am not in any pain thanks to the morphine they have been administering. It is likely that several ribs have been harmed but it is too early to diagnose everything, as my entire body is tensed up. The doctor claims I will be done flying for the next 2 – 3 months, which would be terrible. I hope to be better before too long."[5]

Following his recovery, he returned to flying duty with his previous Feldflieger-Abteilung 6b on

Above: Lt. Clemens Vollert, special duty officer of Jagdstaffel 76, former observer, and close friend of Walter Böning.

September 20, 1916.[6]

On September 2, 1916, during his absence, his Feldflieger-Abteilung had re-located to another area of operations, joining the 1st Army in order to resume missions against enemy positions along the Somme River. Along with Bavarian Artillerieflieger-Abteilung 101 and Kampfstaffel 28, the unit had set up a new base near the small town of Ronssoy. Concerning the situation at the front, he reported home on the next day:

"The skies are alive with aircraft, much like the way to Bloh (a peasantry located to the north-west of Oldenburg, Autor´s note) *with mosquitoes in the summer. One can easily spot 50 – 100 aircraft without the need to turn one's head. Flights pass at altitudes between 500 - 1000 meters in all weather conditions. Hardly any shots are fired by artillery, but there is heavy fire from countless other aircraft. Countless bullets fly through the air, injuring even those not directly engaged in aerial combat. We only fly in Geschwader strength, and we are able to observe just about everything. We really do get things accomplished now, day by day, new batteries are photographed, artillery fire is being directed, etc."*[7]

During his flights along the front, he repeatedly witnessed Jagdstaffel 2 in aerial combat, which he reported to the family back home on September 25, 1916:

"Perhaps you will be interested in hearing about

Above: Ready for take-off; Lt. Clemens Vollert and Lt. Walter Böning with Aviatik C I, Bavarian Feldflieger-Abteilung 6.

Boelckes unit. These Jagdstaffeln and Kampfstaffeln are also required to fly in Geschwader of 6–12 aircraft at all times, as we are, otherwise they would get crushed. They fly at high altitudes, close together. As soon as leader Boelcke raises one arm, it means attention; if he raises both arms, he is signaling a nosedive at an enemy Geschwader. Yesterday, we watched as 7 were shot out of an enemy formation. Boelcke recognizes every single enemy aircraft even from a great distance and knows just how to attack them."[8]

These observations soon woke his desire to become a Jagdflieger (fighter pilot) himself. On October 2, 1916, he seized the opportunity to speak with Hptm. Oswald Boelcke:

"This afternoon, I had a potentially important conversation with Hptm. Boelcke. An Oberleutnant of his Jagdstaffel (Oblt. Kirmaier, who was at Schleißheim at the same time as me) had recommended me to him. They were visiting Bavarian Feldflieger-Abteilung 1 nearby and let me come over for a visit.

Boelcke asked if I felt like joining his Jagdstaffel as a single-seater pilot. The Staffel mostly flies single-seater, biplane fighters. You might have guessed my response: I shook both of his hands and responded that I would be very grateful if he could accomplish to arrange for my transfer. The catch is that there is a mandate issued by the Bavarian War Office, pilots who had been trained in Bavaria were not to be transferred due to the department's own shortage of pilots. Now, I am curious to find out if the almighty Boelcke will succeed in convincing the Bavarians to deviate from their principles (they are notorious for making no exceptions).

Hptm. Hell (Leader of Feldflieger-Abteilung 6b, note of the authors) made rather a sour face, but there is no holding me back under these circumstances. The circumstances are as follows:

1. *Boelckes Staffel is the best of the entire German Army*
2. *Since they fly (inline-engined) biplanes, I will*

Above: Lt. Walter Böning taking up the rather cramped rear seat as the pilot of his Aviativ C I at Bavarian Feldflieger-Abteilung 6. The observer seated in front of him, reigns over a fairly spacious compartment fitted with a huge gun ring mounting two machine guns. Note the metal tubes bent into shape to prevent the observer from accidentally firing through the propeller arc.

not be obliged to re-train to fly monoplanes with rotary engines.

3. *I will not be left to my own devices at the beginning, as the Staffel always flies in a Geschwader formation.*

I certainly find it needless to say that I feel up to the task in spite of rarely having flown single-seater aircraft; as my ultimate goal, I had always envisioned myself flying fighters, and I would consider myself incredibly lucky if my hopes came to pass.[9]

However, his transfer to the single-seater arm of the Fliegertruppe did not progress as quickly as Walter Böning had wished. His commanding officer, Hptm. Hell was not keen at all on letting him go.

On October 30, he wrote somewhat resignedly:

"The outcome of my possible transfer to the Jagdstaffel is questionable at best. Boelcke told me he would put in a new request for my transfer again the next time a pilot was lost. It is a great pity that it had to him in the event. He undisputedly shot down 40 aircraft; there will hardly ever be a second Boelcke."[10]

On November 22, 1916, he was finally able to share some news with his parents:

"I have been chosen to join a newly formed Jagdstaffel..."[11]

Four days later, on November 26, 1916, he was on his way to Cambrai to begin training in the new "fighting machine", as he put it. Two days later he reported that a newly formed Jagdstaffel would be the destination of his transfer. He then traveled to Ronssoy near Epéhy, where Jagdstaffel 19 was formed. Initially, Bavarian Oblt. Franz Walz was the Staffelführer of Jagdstaffel 19. Only a little more than three weeks later, however, Walz was appointed Staffelführer of Jagdstaffel "Boelcke"; Oblt. Erich Hahn, a Saxonian, assumed his position in Jagstaffel 19.

On April 6, 1917, Walter Böning achieved the first aerial victory for Jagdstaffel 19. He described the aircraft as *"single-engined lattice tail"*[12] Due to his leadership skills, Staffelführer Oblt. Erich Hahn soon designated him Kettenführer.[13]

Lt. Erich Hahn fell in battle on September 4, 1917. His successor, Leutnant Ernst Heß wrote the following assessment of Walter Böning:

„Aviation unit: Jagdstaffel 19
Date: 3. October 1917
First and last name: Walter Böning
Patent: promoted to Leutnant d. Res. on October 25, 1915
Peacetime unit: Infantry Reg. Nr.19 (Bavaria)
Present department: pilot
Since: February 10, 1916
Officer professional status while on leave: stud. med. (medical student)
Assessment:
Leutnant d. Res. Boening has been with the Staffel since November 26, 1916. He has certainly proven himself as a soldier, and he has displayed exceptional camaraderie. His loyalty is paired with recklessness and absolute cold-bloodedness which are often required for successful aerial combat, especially against a superior number of enemies. He is an excellent fighter pilot. The skills displayed by him as an aviator are excellent.

He served as deputy Offizier z. bes. Verwendung (special duties officer), consequently he has had insight into the internal services and operations of a Jagdstaffel.

He is well-liked by his comrades and superiors alike due to his reliability, his open nature and tactful behavior. I believe him to possess all the necessary qualities in order to become a successful Jagdstaffelführer.

Heß
Leutnant and Staffelführer"[14]

By the time he was appointed Staffelführer of Jagdstaffel 76 on October 2, 1917, he had already achieved seven aerial victories, one of which had only been reported as "forced to land beyond enemy lines". His achievements made him one of the most successful fighter pilots serving with Jagdstaffel 19 of his time.[15]

When Walter Böning arrived at Schleißheim on October 7, 1917, he was surprised to find out that not all departments there had been made aware of the imminent formation of his Jagdstaffel. In one of the letters to his family, dated October 11, he wrote:

"All I can say is, my head is spinning from all the confusion. Both the Flieger-Ersatz-Abteilung as well as I myself in the field had been given orders to form the Staffel. But higher instances (Königlich Bayerische Inspektion des Militärluftfahrwesens, (Royal Bavarian Inspectorate of Military Aviation), author's note) claim not to have any knowledge of these orders. Nevertheless, I had received direct orders from Kogenluft (Kommandierender General der Luftstreitkräfte, (Commanding officer of the air-force). I will have gathered my entire crew soon, but that will not suffice in order to be able to depart (to the front) on October 15, which I should be, according to what is being expected. Thankfully, I am not only stubborn but also thick-skinned."[16]

As outlined in another letter, the challenges he faced while establishing his Staffel had only just begun:

"First, there is the Prussian War Office which issues all of the orders, but then, there is also the Bavarian War Office that pretends they are in charge of issuing orders. Occasionally, one branch issues orders without previous notification from Munich. I-Luft is in charge of ordering crews and equipment, but they are not doing anything as they had not been officially notified. Particularism is certainly at its peak still.

Well, in the meantime I have managed to convince the good people that I am justified in the requests I have made on behalf of my Staffel. I have gathered all my men with the exception of drivers. The pilots will join me in the field. I can see a ray of hope again as things are moving forward, but it was infuriating at times. It is often my experience that older officers are just as helpless as the youngest ones when it comes to making mistakes or thinking they could have made one. Decisiveness is something I miss greatly. According to local branch orders, I was to receive trucks and personnel from the commander-in-charge once I got to the field. I found this highly suspicious from the beginning, as this would cause major challenges while unloading. My suspicions were confirmed when I went to the inspection office; the local Commander had applied orders which exclusively related to the conversion from Prussian to Bavarian Jagdstaffeln to the formation of my new Jagdstaffel.

I have already made several round trips between Munich and Schleißheim; each time, I have had to inform the Commander (der Flieger-Ersatz-Abteilung, author's note) of things that had been incorrectly ordered and orders which had been incorrectly issued, etc. Of course, one must not be

Lt. Ernst Heß' (Staffelführer of Jagdstaffel 19) assessment of Walter Böning.
OP 38021, Bavarian Main State Archive, Munich.

2

Fliegerverband: Jagdstaffel 19 **Datum:** 3. Oktober 1917.

Vor- und Zuname: Walter Böning

Dienstgrad: Leutnant d. Res.

Patent: 25. 10. 15. zum Leutnant d. Res. befördert.

Friedenstruppenteil: Infanterie Reg. Nr. 19. (bayr.)

Jetzige Dienststellung: Flugzeugführer

Seit wann? 10. 2. 16.

Bei Offizieren des Beurlaubtenstandes Beruf: stud. med.

Beurteilung:

Dienstlich. Besondere Befähigung und Leistungen. Ausserdienstlich. Allgemeine und militärische Bildung und Begabung. Persönlichkeit und Charakter. Für welche Feldverwendung, ob zur Uebernahme in die Fliegertruppe geeignet.

Leutnant d. Res. Boening befindet sich seit dem 26. Novbr. 1916 bei der Staffel. Er hat sich hier durchaus als Soldat, wie in kameradschaftlicher Beziehung bewährt. Pflichttreue, gepaart mit frischem Draufgängertum und großer Kaltblütigkeit, die er oft im siegreichen Luftkampf auch gegen Überzahl zu beweisen Gelegenheit hatte, lassen ihn als ausgezeichneten Jagdflieger erscheinen. Sein rein fliegerisches Können ist ausgezeichnet.

Vertretungsweise war er auch Offizier z. bes. Verwendung, sodaß er auch in den inneren Dienst und die Geschäfte einer Jagdstaffel eingeweiht ist. Durch seine Zuverlässigkeit, sein gerades und offenes Wesen und taktvolles Benehmen ist er sehr beliebt bei seinen Kameraden und Vorgesetzten.

Ich glaube, daß er alle Eigenschaften, die man von einem selbständigen Jagdstaffelführer verlangen muß, in reichem Maße besitzt.

Heß.

Leutnant und Staffelführer

Einverstanden! Leutnant Böning ist mir längere Zeit als [illegible] [illegible] mit sicherem Auftreten bekannt.

Geyer

too direct as one depends on these people doing their duties."[17]

The heartfelt way in which he was greeted by Clemens Vollert, his friend and former observer, was a silver lining. Clemens Vollert had been serving as observers´ instructor at the Lechfeld compound about 60 kilometers west of Munich. Vollert was fully aware of Böning's arrival as the two friends had remained in contact since their days with "6b". He invited Clemens to have dinner at a tavern in Schleißheim. Over a glass of beer, Walter Böning expressed his wish to have his friend join Jagdstaffel 76 as Offizier zur besonderen Verwendung.

Clemens Vollert had been rather bored working as an instructor in Schleißheim and later at the Lechfeld compound; he enthusiastically accepted Bönings offer. The following day, he put in the request to be transferred to Jagdstaffel 76. While he was unable to leave Lager Lechfeld immediately, he began to support Walter Böning while off duty. Fortunately, Clemens Vollert was on good terms with his superior at Lager Lechfeld who silently tolerated his "spare-time work". He even sent him on leave in order to move into the field with the unit. It was not until November 11, 1917, that official transfer orders for Clemens Vollert finally arrived from Schleißheim. By then, Vollert had long been at Habsheim airfield in Alsace with Jagdstaffel 76. Incredibly, the official chain of Command within the Bavarian I-Luft had taken a full month!

Clemens Vollert was born in Leipzig on April 9, 1891. His father, also called Clemens, was an engineer who had been working in Berlin for several months. During their journey back from Berlin to Leipzig, his wife went into labor earlier than anticipated. As a result, Clemens Junior, a true Bavarian, was actually born in Leipzig (Saxony).[18]

He grew up in Munich and joined the 1st Bavarian Infanterie-Regiment as a one-year volunteer. He was promoted to Gefreiter on April 1, 1914, and to Unteroffizier on July 1, 1914.

At the beginning of the war, he was a member of the 1st Bavarian Infanterie-Regiment. On August 20, 1915, his left lower leg was injured by shrapnel; in addition, he suffered a fractured rib on the left side of his body. Following successful treatment, he was transferred to Ersatz Bataillon 1.bIR on January 20, 1915. On May 12, 1915, he was promoted to Leutnant der Reserve. The same year, he volunteered to join the Fliegertruppe. On February 20, 1916, he was commanded to the Bavarian Flieger-Ersatz-Abteilung in Schleißheim in order to attend observer training, his transfer there came into effect on April 8, 1916.

On April 22, 1916, he arrived at Bavarian Feldflieger-Abteilung 6 via Army Aviation Supply Depot A, where he encountered Walter Böning. Following successful missions, some of which he completed with Walter Böning, he was awarded the observer's badge on August 2, 1916, and the EK I (Iron Cross 1st Class) on December 24, 1916. On July 11, 1917 he was transferred to Beobachterschule 1 in Schleißheim. From there, he was transferred to Fliegerschützenschule at Lager Lechfeld on September 2, 1916.[19]

In addition to all the bureaucratic troubles, Walter Böning realized that the atmosphere had changed significantly since the days when he attended pilot training one-and-a-half years earlier.

The spirit of optimism, the dedication and the passion for aviation were not the same as before. He had met a few young pilots who enthusiastically and eagerly awaited flying and looked forward to being deployed to the front. Later, he even attempted to recruit some of them for service in his Jagdstaffel.

However, he was deeply frustrated by the leadership of the Flieger-Ersatz-Abteilung. It appeared as though Schleißheim was overwhelmed with the task of putting together a new Jagdstaffel. Additionally, he was faced with an extraordinarily slow bureaucratic system which was more concerned with meticulous adherence to official channels and cumbersome requirements instead of assuring the frontline serviceability of a newly created Jagdstaffel.

Support from the Head of Flieger-Ersatz-Abteilung left a lot to be desired. Walter even purchased the dishware for the officer's Kasino in Munich himself, but only after the deputy head of the Flieger-Ersatz-Abteilung had finally agreed to provide a budget for the purchase, following extensive negotiations.

Böning aired his disappointment in writing:

"... in my report I will not be able to avoid pointing out exactly what has been messed up, and everything that might have been accomplished by using the most simple means and proper leadership from above. I will list the ones who have not fulfilled their duties and obligations; this list includes adjutants of the Department (FEA Schleißheim) who, relieved of frontline duties due to anxiety and nervousness to be used elsewhere in expensive times, and who in turn get annoyed by young Officers who move into the field as Staffelführer. These people, who are mostly observers, seem to think that Staffelführer and Flieger-Abteilungs-Führer alike are not required to do much flying, and that they, in short, can just sit back and enjoy the war. One of these people even

remarked to me that now it would be my turn to sit back and relax and laze around. I responded by expressing my unmistakable opinion for which I will likely face small repercussions. Ever since that day, however, I have been back to feeling excellent…"[20]

Incidentially, there were no repercussions for the time Walter Böning lost his cool at the office of a senior officer; it was well-known that they would soon be rid of the young "perpetrator" again.

In the conversation he added the following comment:

"While soldiers were dying out in the field, people in the office refused to process applications and return them due to formal errors. Everything seemed to take forever, and one was under the impression that having to do any type of work was causing the highest inconvenience of them all for these officers. I was obliged to pursue each and every one of my requirements personally; I was up from dusk till dawn in order to be successful in forming the Staffel. Also, the support lent by Clemens' (Vollert, author's note) helped greatly. He knew the personnel there and had a great relationship with the NCO´s and soldiers (from his time in Schleißheim, author's note). I was even able to send the crew on leave before we deployed."[21]

One of the issues Walter Böning was faced with was theft, particularly of leather goods, which were sold on the black market. This was something that had been unthinkable even two years prior. Fortunately, and with some help from Clemens Vollert, Walter Böning managed to recruit a capable Feldwebel who, in turn, knew a handful of Unteroffiziere that were tired of the "pigsty that was Schleißheim", as they termed it. They were happy to get transferred to Jagdstaffel 76. One of their first actions was to organize guards for all equipment and material, which quickly put an end to most of the thefts.

The ground personnel that had been assigned to him proved to be an even greater challenge. They were largely inexperienced and poorly trained. With the help of the acting Feldwebel and other NCO´s, he located a few capable aircraft and engine mechanics for his Jagdstaffel and immediately recruited them for his unit. This achievement resulted in renewed reprimands by the head of the Flieger-Ersatz-Abteilung due to non-compliance with the required chain of command. By then, however, Böning had become completely indifferent to that. After much back and forth, he was finally granted the personnel he had requested. Aircraft mechanics began training ground personnel and assigning the crews according to their abilities and interests.[22]

The bureaucracy he dealt with in Schleißheim was not the only challenge he was faced with, however. Requests sent by the Bavarian Inspectorate of Military Aviation were processed only slowly and with hesitation by the Inspectorate of Military Aviation in Berlin. According to Walter Böning, one was under the impression that these two authorities were working against each, rather than collaborating in the way they should. In a letter to his parents from October 27, he remarked:

"October 27, 1917. The date of my departure remains unknown as the equipment requested from Berlin has taken 14 days to be shipped so far, and still hasn´t arrived yet…I will thank God when I will finally be able to fly again. I will not write for a little while as my anger is running right into my fingers…"

Furthermore, he pointed out:

"Initially, my Staffel will likely be sent to a rather quiet spot at the front, which will be fine by me, as long as I will end up in Flanders next spring. Flanders remains the answer to all German Jagdfliegers' dreams, even if the losses will be greater there."[23]

On November 1, 1917, the Staffel was finally ready to get on the way; Walter, as he explicitly noted, crossed himself three times. After all of the Staffel equipment had been loaded onto trains, the head of the Flieger-Ersatz-Abteilung arrived to officially bid the Staffel farewell. It was, in the memory of Walter Böning, a pathetic rant which was literally dripping with patriotism, fighting spirit, and self-sacrifice.

Following his speech, he leaned over to Walter Böning and said:

"I am very pleased indeed that, in spite of the great difficulties you faced, you managed to successfully form the Staffel on your own devices and put it on the right tracks."[24]

Walter Böning nearly lost his cool at that remark; but then, he swallowed his anger and dutifully thanked him for the support. He was just unbelievably relieved to finally be able to leave Schleißheim behind.

2 Baptism of Fire in Alsace

Armee-Abteilung B, November 1917 – January 1918

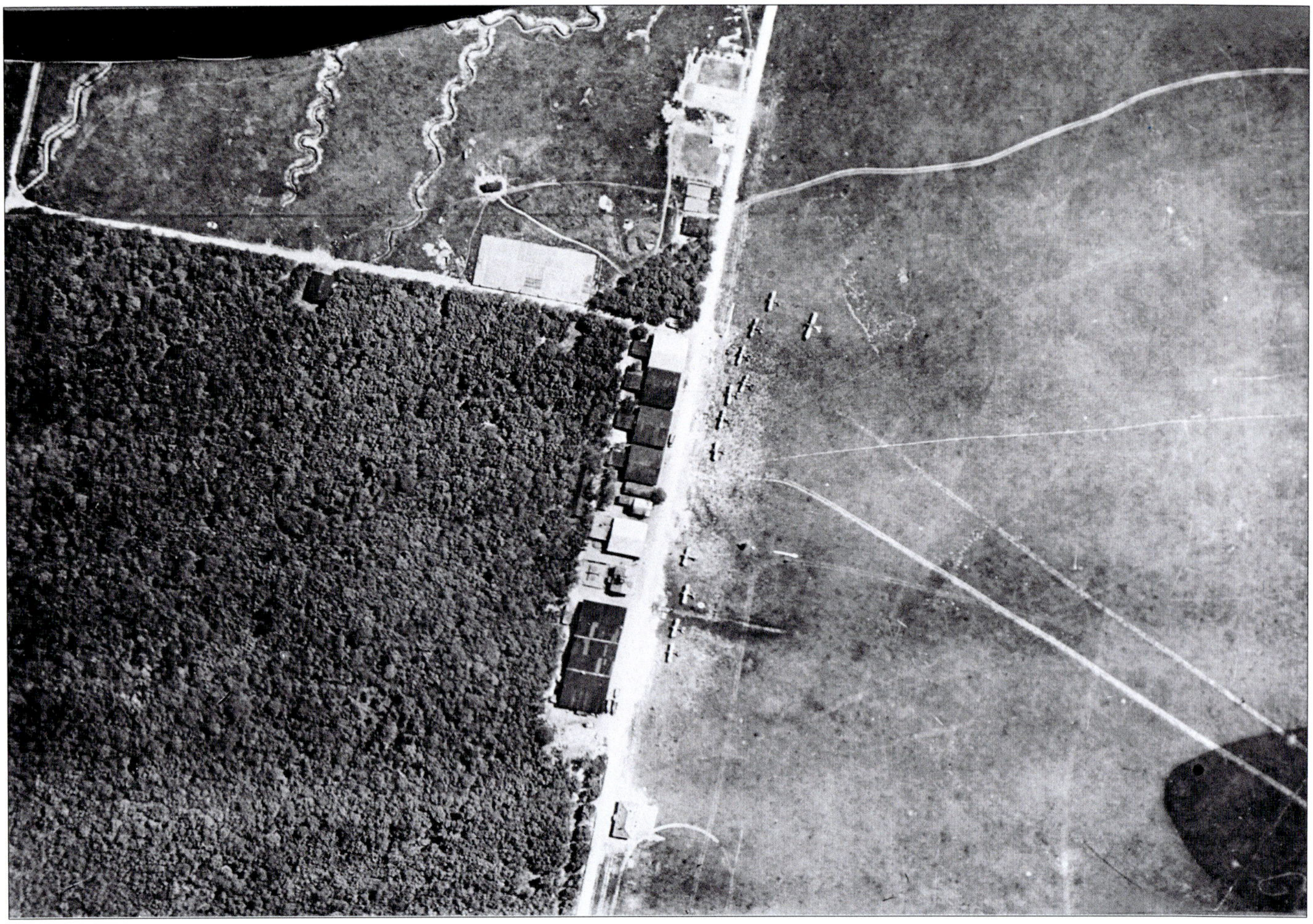

Above: Birds-eye view of the first frontline airfield occupied by Jasta 76 at Habsheim. Near the edge of the forest, the big wooden hangar buildings can be seen. These will be seen in the background of several photos in this book. 12 Albatros single-seat fighter aircraft can are standing in front of the hangar, as well as the squadron "hack" two-seater, which is the biggest aircraft in this view. Note the trenches in the upper left corner of the photo.

On November 4, 1917, the Staffel's rail transport arrived at Mulhouse in the Upper Alsace region. The Staffel set up their quarters at Habsheim airfield in the region of Army Department B. Eventually, the aircraft destined to be used by the Staffel also arrived there.

The area of Army Department B was located in the southern Vosges region. The frontline began close to the Swiss border near the village of Pfetterhausen (Alsatian: Pfatterhüsa) on the French side, and Moos on the German side. Resembling a wavy line, the front extended to the north where it bulged slightly to the east near the Rhône-Canal near the village of Eglingen. The frontline continued its curvy path along a line between Altkirch and Thann, ending to the north abeam Strasbourg, where Army Department A took over responsibilities.

Ever since the tough battles and high losses at the beginning of the war, this area of the front had been fairly calm; German and French troops were opposing each other in carefully constructed shelters, no major fighting took place on the ground.

Habsheim airfield was located in the Rhine valley, approximately 2 km south-east of Mulhouse. It had been in use long before the war and was well-developed; it included hangars, workshops,

Above: Lt. Walter Böning in the cockpit of his Albatros D Va. A telescopic gunsight has been mounted in front of the cockpit offset to the right, and a rear-view mirror is fitted to the centre of the upper wing cut out. His personal insignia, the blue-white-blue fuselage band is visible behind the pilot's seat.

administrative buildings, depots, warehouses, as well as barracks for pilots and ground personnel. Before the war, the Aviatik aircraft factory was located at the edge of the airfield. Due to the close proximity to the front, the company was re-located to Leipzig in 1916, while the now-empty buildings served well as hangars for the units stationed at Habsheim. The take-off and landing conditions at the airfield were favorable, with the exception of wind shear which caused some challenges during the fall and in the spring. Near the Swiss border to the south of the operational area lie the foothills of the Jura mountain range; to the west, the Rhine valley is bordered by the Vosges region.

The Staffel operational area included the Swiss border to the south and Münster (near Colmar) to the north. The front along the line between Niederaspach and Oberburnhaupt was located

Above: Signed photo of Uffz. Siegfried Walther posing in front of his Albatros. He was the Staffel's second most successful fighter pilot behind Walter Böning. (R. Kastner)

approximately 18 km from the airfield and could be reached in about 10 minutes of flying time. The missions largely took place above the Vosges mountain range. This location imposed some challenges, particularly in the case of potential emergency landings, which were difficult to execute successfully in mountainous terrain.

The opponents of the German Fliegertruppe were exclusively units of the French *Aéronautique Militaire* whose approach to aerial combat was to take on a clearly defensive stance. Their primary task was the protection of their own airspace from German intrusions. French aircraft only entered German airspace in order to carry out bombing attacks or reconnaissance missions. French pilots were considered to be generally well-trained opponents. They were dangerous and not to be underestimated.

As opposed to the ground troops, there had been a number of heavy aerial battles in the past.

Above: Signed portrait photo of Lt. Wilhelm Kolb, who was a native of the city of Munich.

Jagdstaffel 41, which had been established at the Habsheim airfield in June of 1917, had accounted for 17 confirmed aerial victories by the end of December 1917. It was during that period of time that Oblt. Maximilian Ziegler named Stege, their first Staffelführer, had fallen in aerial combat over Ensisheim. Vzfw. Rosenfeld had been taken captive by the French following a successful captive balloon attack.[1] In addition to previously mentioned Jagdstaffel 41, the Bavarian Fliegerabteilung (A) 289 was also stationed at the airfield.

After his arrival there, Walter Böning wrote to his parents:

"I arrived at my destination at 2 am. The transport, as well as loading and unloading, were challenging at times; fortunately, we managed better than expected. Most of the people here are 18-19 years old and untrained. Most of them have never been at war; they run around like little children and possess no sense of punctuality, working with them is not always easy. Thanks to the support a number of fairly proficient NCO's which I had taken on charge a time ago, things went fairly smoothly. I have the names of all my pilots, but none of them have arrived in person. However,

Above: Lt. Hans Böhning while serving as an artillery soldier before his transfer to the Fliegertruppe.

Above: Lt. Erwin Wenig in very informal attire as a member of Jagdstaffel 28 in the summer of 1917.

that´s not an issue for the time being since we´ve been having some dense fog for days."[2]

Walter Böning was more than puzzled when he learned that the local airfield command had not set up any hangars or living quarters for him and his Staffel. In response to the Staffelführer's inquiry, the airfield commander responded that they had not been issued any corresponding directives from Munich. Walter Böning shook his head and proceeded to meet with Lt. Georg Schlenker, Staffelführer of Jagdstaffel 41. Schlenker was just as puzzled; he had no knowledge of any transfer orders or news of being replaced by Jagdstaffel 76. He graciously offered to assist Walter Böning in finding living quarters for his Staffel. He also advised his Staffel to make as much room as possible in their hangars and living quarters in order to share what was available with the Bavarian "76´ers".

The airfield premises were now so crowded that Jagdstaffel 76 was obliged to store equipment outside of the airfield boundaries, which led to the theft of items yet again. Walter Böning set up guards which curbed some of the theft but did not eliminate the problem completely. He was under the impression that things had gone from bad to worse:

"My main concern during the present state of emergency remains the accommodation of the Staffel. We came to replace a local Staffel but now we are living in makeshift accommodations. The crews, offices, etc. have been relocated to a town 30 minutes from here; machines and technical operations have been consolidated with the other Staffel's equipment. Consequently, items frequently disappear; cockpit instruments, tools, and many other useful items. It is impossible to set up enough guards to keep a watchful eye on everything."[3]

Off duty, the Alsace region was a place filled with sleepy, medieval villages and romantic castles which made it easy to forget that there was a war going on. The food supply situation remained intact, there were plenty of taverns with an abundance of good food, and wine cellars were filled with great quality wines. As Alsace had been part of the German Reich again since 1871, the pilots found themselves at home, rather than being stationed in an occupied

Above: Lt. Xaver Dannhuber pictured during his initial flight training at the Bavarian "Flieger-Ersatz-Abteilung 1 at Schleißheim, holding the inevitable cigarette in his right hand. (A. Imrie)

Above: Due to his extensive combat experience, Karolus Bärenfänger, pictured here in the rank of Vizefeldwebel during his earlier service with Jagdstaffel 28, became one the backbone members of the newly formed Staffel.

territory such as France or Belgium.

Joachim von Hippel was a member of Jagdstaffel 71. On February 17, 1918, his Staffel was transferred to Habsheim, immediately following its formation at Flieger-Ersatz-Abteilung 13 in Bromberg. The Staffel remained operational at Habsheim until the end of the war. He reported that in the Alsace, one would get the overall impression that one was taking part in a military maneuver, had it not been for their daily flights to the front, which regularly resulted in aerial combats with formations of French flyers.[4]

On November 5, 1917, the first two pilots from Jagdstaffelschule II in Nivelles, Lt. Wilhelm Kolb and Unteroffizier Siegfried Walther, arrived at Jagdstaffel 76.

Leutnant Wilhelm Kolb was born in Munich on June 18, 1895. At the outbreak of the war, he deployed as Fähnrich with the 5th battery of the 7th Bavarian Field-Artillery-Regiment. On February 24, 1915, Kolb was promoted to Leutnant. His application to join the Fliegertruppe was accepted, allowing him to begin pilot training after receiving orders to join Flieger-Ersatz-Abteilung 1 in Schleißheim on February 15, 1917. He first completed his technical training in Milbertshofen, followed by practical flight training at Militär-Fliegerschule 4 at Lager Lechfeld. Leutnant Kolb served as a pilot with Flieger-Abteilung (A) 297 from August 20, 1917, until October 11, 1917.[5]

Walter Böning described Wilhelm Kolb as follows: *"...one of the most talented pilots I have ever experienced, besides Böhning, a naturally gifted flyer; in addition, he is one of the most likeable and friendly comrades I have ever known. He was always happy and in a good mood, and thoroughly reliable during missions. His untimely death due to wing failure ended a great aviator´ future."*[6]

Siegfried Walther was born on April 25, 1896, in

Raguhn near Dessau. He worked as a photographer in Zerbst and joined the Bavarian Army when the war began and moved into the field in December 1914 as part of Reserve-Infanterie-Regiment 16. He joined the Fliegertruppe at the end of 1915 and was transferred to Schleißheim in December of 1915. He arrived at Army aviation supply depot "von Stratz" on February 22, 1916, in order to serve as the assistant to Leutnant Rudolf Burger. On March 12, 1916, he was transferred along with Leutnant Burger to Bavarian Feld-Flieger-Abteilung 1. On June 28, 1916, Siegfied Walther was ordered back to FEA Schleißheim in order to begin pilot training. He completed technical training at Motorenschule Milbertshofen near Munich in November 1916, followed by practical flight training at Militär-Fliegerschule 3 in Fürth and Fliegerschule 5 in Gersthofen near Augsburg. Upon completion of pilot training, he was transferred to Armee-Flugpark 7 on May 16, 1917 and assigned to Bavarian Schutzstaffel 22. He served with the Schutzstaffel at the front until October 13, 1917.[7]

Walter Böning later described him as follows:
"...a good, reliable Jagdflieger who sometimes was a bit too impetuous and reckless. At times, he had to be kept in check if he was about to take an unnecessary risk. In any case, he was a reinforcement for the Staffel."[8]

The following day, two additional pilots arrived at the Staffel, Leutnant der Reserve Hans Böhning from Jagdstaffel 36, and Leutnant Erwin Wenig from Jagdstaffel 28. As mentioned previously, Walter Böning had requested Hans Böhning's transfer to his new Staffel while he was still serving with Jagdstaffel 19.

Hans Böhning was born on July 6, 1893 in Grottau near Reichenberg in Sudentenland (today Hrádek nad Nisou, Czech Republic). In 1907, his family moved to Bremen where he attended Gymnasium (secondary school). There, he saw flying machines for the first time – they immediately captivated his interest. Upon graduating from Gymnasium, he traveled to Johannisthal in 1912. Johannisthal was considered the core of German aviation, and Böhning was determined to attend "Aviator" training. He went on his first flights with aviation pioneer Felix Laitsch. Lacking money for flight training, he worked as a helper at the airfield in order to earn occasional flying lessons. Against the will of his father, who showed little understanding for the "nonsense" his son was interested in, Böhning had joined an organization called "Nord-West-Deutscher-Flugwettbewerb 1912 (North-West German Flying Competition, note of the authors)". Lacking the financial means to support his dreams, he was soon forced to follow his father's wishes and proceed with "proper" studies. He enrolled at the University in Erlangen in the Kingdom of Bavaria to begin his studies in philosophy and archaeology.[9]

Above: Vzfw. Paul Wagner as a member of the 4th Bavarian Chevauleger-Regiment.

Much like Walter Böning, he enjoyed the Bavarian way of life; he soon felt like a Bavarian himself. He vividly expressed this by having white and blue lozenges painted on his aircraft as his personal insignia while serving with Jagdstaffel 36. When the war began, he was in his fourth semester; he signed up as a war volunteer along with thousands of other students at the time. He traveled to Schleißheim to apply for service with the Royal Bavarian Fliegertruppe. However, he soon had to learn that, due to overwhelming interest, only licensed pilots were accepted into the Fliegertruppe for the time being.[10] Accordingly, he decided to join the 10th Royal Bavarian Field-Artillery-Regiment on August 14, 1914 and joined the fighting. Following his promotion to Leutnant der Reserve on November 28, 1915 he suffered injury to his knee by a hand grenade on June 6, 1916, which resulted in pain while walking. This condition made it impossible for

Above: Jagdstaffel 76 at Habsheim airfield in the spring of 1918. Pictured from the left are: Lt. Hans Böhning, Uffz. Paul Wagner, Lt. Walter Böning, Lt. Clemens Vollert Lt. Carolus Bärenfänger and Uffz. Siegfried Walther.

him to continue serving with the field artillery, so he re-applied to join the Fliegertruppe.

On October 1, 1916, he received orders to begin pilot training with Bavarian Flieger-Ersatz-Abteilung 1 in Schleißheim. Following graduation from pilot training at Bavarian Militär-Fliegerschule 4 at Lager Lechfeld, Böhning was transferred via Armee-Flug-Park 3 to Bavarian Flieger Abteilung (A) 290 on April 26, 1917. His duties as Artillerieflieger hardly coincided with his temperament; soon, he applied for transfer to the fighter units. On June 18, 1917 he was transferred to Armee-Flugpark 4. Upon completion of fighter pilot training, he was transferred to Royal Prussian Jagdstaffel 36 just two weeks later, on July 3, 1917.

Already on the following day, he was shot down during aerial combat and suffered a nasal fracture in the resulting emergency landing, which could be treated locally.

Until his transfer to Jagdstaffel 76 on November 6, 1917 he had achieved four confirmed aerial victories. He was going to serve as technical officer at Jasta 76 following his transfer to the unit; this meant that, in addition to his pilot duties, he would supervise the aircraft's technical status and the replenishment of technical supplies. (11)

In a letter to his parents, Walter Böning wrote:

"I had requested several men personally…a technical officer (Lt. Böhning from Bremen), with whom I have been friends with for a long time…" (12)

Furthermore, he remembered Hans Böhning as follows:

"He had a constant thirst for action, was highly intelligent, very technically interested, and above all, he was absolutely reliable. He took care of all technical aspects of the Staffel and supported me in instructing the pilots. He was the best fighter pilot of the Staffel, and a great friend. Unfortunately, he was only with me for a short period of time before he was assigned his own Staffel."(13)

Lt. Erwin Wenig was another one of the pilots personally requested by Walter Böning. He was born on July 7, 1893 in the small town of Weiler in the administrative district of Neuburg in Swabia. Due to his technical interests, he had joined the 1st Bavarian Telegraph-Bataillon as an officer candidate before the war; by the outbreak of the war, he had already been promoted to Leutnant. Once again, he was motivated by his technical interests when he applied for transfer to the Fliegertruppe in mid-September 1914.

On September 21, 1914 he was transferred to Schleißheim in order to begin training as an observer with Bavarian Flieger-Ersatz-Abteilung. On January 15, 1915 he joined Bavarian FeldfliegerAbteilung

Above: Jagdstaffel 76, from left: Lt. Hans Böhning, Uffz. Paul Wagner, Lt. Walter Böning, Lt. Clemens Vollert, Lt. Carolus Bärenfänger.

7, and less than a week later he was transferred to Schleißheim to join Bavarian Feldflieger-Abteilung 8. This unit was still in the process of being formed, and at the time consisted of only four air crew members, Leutnant Otto Deßloch, Leutnant August Dittmar, as well as Unteroffiziere Otto Kissenberth, and Gottfried Ruhl.

During one of many combat missions, Erwin Wenig was involved in a heated aerial battle on March 21, 1915. In the course of this, his pilot, Vizefeldwebel Kissenberth sustained severe injuries to his thigh. Following their emergency landing, he was immediately transported to the field hospital in Colmar.[14]

In time, Erwin Wenig realized that he was not destined to fly as an observer. He subsequently applied for pilot training; his request was granted, and he was transferred to Schleißheim on July 13, 1915. On October 12, 1915, he returned to Bavarian Feldflieger-Abteilung 8 as a pilot. In January 1916, he arrived at the Kampfeinsitzer-Kommando (single-seater fighter Command) Ensisheim to fly Fokker E II and Pfalz E II.

From July 1, 1916 he was given orders to join the KampfeinsitzerAbteilung des Großen Hauptquartiers (single-seater fighter detachment for the Great Headquarter) for one month before returning to his Kampf-Einsitzer-Kommando at Ensisheim. Later, this Kampeinsitzer-Kommando would provide the nucleus personnel and equipment for the formation of Jagdstaffel 16. On April 26, 1917, Erwin Wenig was temporarily commanded (which was changed into a permanent order of transfer one week later) to Flanders to join Jagdstaffel 28, where he achieved his first aerial victory. With the addition of Erwin Wenig, Jagdstaffel 76 had gained another very experienced fighter pilot.[15]

Max Speidel, his special duties officer at Jagdstaffel 80, remembered him as an exceptional superior who knew how to lead his subordinates. He was friendly and motivating; in modern parlance, the term "team player" would describe him perfectly. As Kettenführer with Jagdstaffeln 28 and 76, he was known for being a prudent fighter pilot who, in spite of his fighting spirit, always thought of the safety of his pilots first.

Left: Lt. Georg Schlenker, Staffelführer of Jagdstaffel 41 at the airfield at Habsheim. Both units shared the airfield until Jasta 41 was transferred to Phalempin on 19/20 March 1918.

He was a relaxed and sociable Swabian, both on and off duty. He was a hobby photographer and would turn this hobby into his profession after the war.[16]

By November 9, 1917, the first Albatros aircraft had arrived at the Staffel. On that day, Walter Böning wrote a letter home from Habsheim:

"I have gained quite a bit of momentum; with the support of the special duties officer I am hoping to be able to reduce my office duties to the occasional signature, as I would much prefer to be flying. As far as I can tell at the moment, enemy flight activity is slightly higher than in the Champagne region. For the first time in a while, I heard cannon thunder; in order to round off the warlike image of the situation, several aircraft were dispatched on both sides. The Staffel that I was sent to replace here (Jagdstaffel 41, author's note) had a number of successes. We were not yet part of them, but as of today I´m ready for the front.

The food supply situation has been resolved by now, along with the replenishment of materials. When one is transferred into a new Army area, one has to make many new acquaintances with people working at the commissariat; if I had not done this, I would have been sitting in an unheated office, typing up requests for food and materials with frostbitten fingers in order to get what could have been here 14 days earlier."[17]

Pilots Leutnant der Reserve Franz-Xaver Danhuber, Vizefeldwebel Karolus Bärenfänger, and Vizefeldwebel Paul Wagner arrived at the Staffel on November 15, 1917. At this point, the Staffel had eight pilots at its disposal.

Franz-Xaver Danhuber was born on January 27, 1891 in Gars am Inn, north of Wasserburg. He had already completed military service in 1911 by volunteering for service with the Bavarian 2nd Infanterie Regiment for a one-year period. In 1914, he deployed with the 1st Company of the Bavarian 1st Reserve-Infantry-Regiment. In the spring of 1916, he forwarded his request to be transferred to the Fliegertruppe in order to begin pilot training. On May 6, 1916, he was transferred to Bavarian Flieger-

Facing Page, Below: The same wreck seen from a different perspective after the accident. The blue-white unit marking on the horizontal tail surfaces shows well in these two photos. The lack of visible stains on the airframe suggests that the airframe was still fairly new at the time of the crash. (G. VanWyngarden)

Above: This Albatros D III (O.A.W.) was actually the personal plane of Lt. Xaver Danhuber. It is seen here after it was crashed by Uffz. Walther on November 23, 1917. (R.Absmeier)

Above: Mission briefing at Habsheim airfield, from left: Uffz. Wagner, (concealed), (concealed), Lt. Kolb, Lt. Böning, paymaster Schweppech, unknown, Lt. Dietlen, Jagdstaffel 41. An Albatros upper wing can be seen at left, stored on its leading edge in the hangar door.

Facing Page, Above: Some aircraft only enjoyed very short careers without being subjected to enemy fire. This and the following two photos document the crash landing by Uffz. Walther with his Albatros D III (O.A.W.) at Habsheim airfield, in early December 1917. (R. Absmeier)

Facing Page, Below: A different perspective of the wreck of Albatros D III (O.A.W.) crashed by Uffz. Walther. The fuselage, fin and ruder of this plane have already been painted in his personal color, red. The fuselage bottom is left in plain factory finish, while the unit marking has not yet been applied to the horizontal tail surfaces.

Ersatz-Abteilung in Schleißheim. On October 31, 1916, he joined Bavarian Kampfstaffel 35, serving as a pilot in Kampfgeschwader 6 (renamed Schutzstaffel 26 on February 12, 1917).

Six months later, he requested to be transferred to a fighter unit. On May 8, 1917, he received orders to attend conversion training at Jagdstaffelschule I in Valenciennes, where he successfully completed the single-seater course. While he attended the course, he was also promoted to the rank of Leutnant der Reserve. On May 25, 1917, his transfer to Jagdstaffel 26 followed via Armee-Flugpark 2. His new Jasta commander, Oblt. Bruno Loerzer, was well-known for personally educating and training all of his pilots. Franz-Xaver Danhuber was a very experienced pilot.

On August 12, 1916 he achieved his first aerial victory; until his injury on October 18, 1917, he added nine additional aerial victories to his tally. By the time he was transferred to Jagdstaffel 76, he had achieved 10 aerial victories and was the Staffel's most successful pilot.[18]

The two other newcomers were close acquaintances of Erwin Wenig, as they both hailed from his former Royal Württemberg Jagdstaffel

Above: By the time Vzfw. Siegfried Walther crashed this Albatros D III (O.A.W.), no unit markings had yet been applied to the nose of the aircraft either. This indicates that the plane was lost after one of some test flights. The military number of the plane is still present on the prop spinner but is too indistinct to make out clearly in the photo.

Above: Another case of one man doing the work while the others are relaxing. Albatros D Va D.5765/17 has its radiator and engine checked following its delivery to the Staffel, note the water and oil cans on the ground. In the background, a group of comrades are taking a nap on a bench in the winter sunshine. This aircraft later became the personal mount of Lt. Bärenfänger who had it marked with white and blue lengthwise stripes. Lt. Max Schick was taken captive by the French after parachuting from this aircraft on June 23, 1918. (R. Absmeier)

28. They were both members of the Bavarian Fliegertruppe, and their transfer to a Bavarian Jagdstaffel had already been pending. Therefore, Erwin Wenig had suggested that they should arrange for a joint transfer to Jagdstaffel 76 along with him.

Vzfw. Karolus Bärenfänger was born in Munich on December 8, 1892. Following the outbreak of the war, he joined the Bavarian Kraftfahrer-Ersatz-Abteilung as a war volunteer. On February 7, 1915 he was promoted to Unteroffizier and Offizier-Aspirant. In the spring of 1915, he was transferred to Flieger-Ersatz-Abteilung 4 in Posen. He completed his pilot training on June 27, 1916 by successfully passing the third exam.

On August 17, 1916 he was transferred to Flieger-Ersatz-Abteilung 7 in Cologne. From there, he joined Kampf-Einsitzer Staffel (KeSt) 6 in Bonn-Hangelar which was tasked with homeland defense duties. Following a stopover at Übungspark Ost, he transferred to Kampf-Einsitzer Staffel 2 (later renamed KeSt 1) in Mannheim on October 17, 1916. Bärenfänger was transferred to FEA Schleißheim at the beginning of December 1916. On January 4, 1917 he joined Jagdstaffel 6 via Armee-Flug-Park 2 before he transferred to Jagdstaffel 28 on January 24, 1917. There, he achieved his first confirmed aerial victory over a Sopwith biplane on May 25, 1917. He too was considered an experienced and prudent fighter pilot.[19]

Vzfw. Paul Wagner hailed from Augsburg; he was born on February 8, 1895. Before the war, he studied to become an engineer in Berlin. In September 1914, he went to war as a member of the 4th Bavarian Chevauleger-Regiment König (light cavalry). He volunteered to join the Fliegertruppe in the spring of 1916 and was transferred to attend pilot training in Schleißheim on June 4, 1916. On July 15, 1916 he was transferred to Bavarian Kampfstaffel 31 of Kampfgeschwader 6 (renamed Bavarian Schutzstaffel 22 on February 12, 1917). On April 12, 1917 he was commanded to Jagdstaffel 28, which turned into a permanent transfer on June 1, 1917. He was yet to score his initial aerial victory. Walter Böning recalled him to be a very proud cavalryman and member of the Chevauleger.[20]

In due course, the Staffel was busy flying in their new aircraft which had been issued to them by Armee-Flugpark B. They also spent time familiarizing themselves with the geographical aspects of the operational areas. The fact that the Staffelführer (from his time with Feldflieger-Abteilung 6b) and Erwin Wenig (from his time with Feldflieger-Abteilung 8b) were already well familiar with the area from their time spent with Kampfeinsitzer-Kommando Ensisheim and Jagdstaffel 16 turned out to be very helpful indeed.

He was able to provide his fellow pilots with exact instructions regarding points of reference, flight routes and information about the ever-

Above: Lt. Wilhelm Kolb and Lt. Hans Böhning on a snow-covered Habsheim airfield in front of Lt. Bärenfänger's white and blue striped Albatros D III (O.A.W.). The aircraft has been fitted with a telescopic gunsight and a rear-view mirror, arranged in identical fashion to Lt. Walter Böning´s aircraft.

Above: Wilhelm Kolb with three falcon chicks, he found abandoned and which he raised by himself.

changing wind conditions in the Vosges region. The railroad tracks leading north-west from Mülheim to the front in Sennheim, as well as northbound towards Colmar, were considered to be the most important landmarks. Another point of reference was the Rhine-Rhône Canal leading south-west from Mülheim.

On November 23, 1917 a serious accident occurred. Unteroffizier Eduard Feig, a member of the ground crew, was caught by Uffz. Siegfried Walther's aircraft, which was taking off at 10:30; Feig was hurled into the propeller circle. He sustained grave internal injuries and died during the transport to field hospital 36. The aircraft was a total write-off. The pilot was deeply shaken by the accident but did not sustain any injuries himself.

On November 26, 1917, they encountered French fighter aircraft in the air for the very first time. Lt. Walter Böning departed to the front alongside his most experienced pilots, Lt. Hans Böhning and Lt. Erwin Wenig. They spotted five French aircraft, described as Nieuports, which were flying approximately 200–300 meters above them. The German pilots pulled up their aircraft and shot at

Facing Page, Top: The sad wreckage of Lt. Wilhelm Kolb's crashed Albatros D Va D.5294/17. The fuselage of the aircraft was painted "baby blue", and the yellow and black fuselage band was a reference to the colors of his hometown Munich. The rudder was a replacement component which had not been painted at the time of the crash. The upper surfaces of the wings carried a painted camouflage scheme while the lower surfaces were covered in five-color dyed fabric.

Facing Page, Middle: The crashed Albatros D Va D.5294/17 from the other side. The impact was so severe that the landing gear was torn out of the wooden fuselage bottom, yet it remained in one piece. The tonal difference between the "baby blue" fuselage and the "Bavarian blue" horizontal tailplane is evident in this view.

Facing Page, Bottom: Both lower wings broke off the fuselage at the wing roots and ended up lying on top of each other after the crash. The upper wing broke into several pieces as a result of the crash.

Above: Uffz. Karl Hopf from Würzburg joined Jagdstaffel 76 on the date of Lt. Wilhelm Kolb´s fatal crash.

the Nieuports from below. They did not manage to reach the French aircraft's altitude in time before they turned away and finally disappeared behind the French front.

Walter Böning expressed his satisfaction with his crew in a letter he wrote home on November 27, 1917:

"For the time being, we´re subjected to a persistent period of rain. In the valley between three mountain ranges we can only fly when northerly winds prevail. To be honest, I was fine with that at the beginning, as this allowed some additional time to align the armament of the aircraft and have them set up properly, and getting this work done by partly completely inexperienced men takes time. As far as I can judge so far, I have been very fortunate to have

Above: Winter has arrived at Habsheim, and the trees behind the hangars are covered with a thick layer of freshly fallen snow.

such great pilots; they are exceptional aviators, and one could not ask for better comrades than them. Oddly enough, besides myself, hardly any of them appear to play cards; but they all have incredible musical talents, and I admit I am a bit jealous of that. Two of the men play the guitar, another two

Above: The snow-covered Jagdstaffel 76 officer's casino glistens in the winter sun.

play the mandolin, one plays the piano, and another plays the violin… our evenings are usually pretty entertaining."[21]

While the organizational aspects of his Staffel were proceeding well, had found out that his initial poor impressions of the Armee-Abteilung B military administration were already confirmed at this early stage. He continued in the same letter:

"You might think that there is nothing left to desire but there are a great many things that have not been going as I had hoped. I have been dealing with a deputy Kofl (Kommandeur der Flieger der Armee-Abteilung-B) who is a pedant and a stickler to the worst degree. It is normal for paperwork to contain small mistakes initially. However, he returns urgent paperwork with the request to re-write them because there is a word which might be considered redundant to the reader… material replenishment has been another constant trouble; we either receive a small share of the equipment urgently needed, or they will ship substitutions or nothing that could be of any use at all."[22]

In the conversation, Walter Böning added that according to his impression, Fliegeroffiziere who had turnend out to be simply inept during their service with other Armies, were relegated to Armee-Abteilung B. Here they were expected to cause the least harm and not get themselves into as much trouble. As a consequence, the units stationed there now had to endure their "talents". There was practically no real coordination of flight schedules by the Kommandeur der Flieger, as had been the case for Jagdstaffel 19 in the German 7th and 1st Armies. It was left up to each individual Staffelführer to arrange for the flight schedules of his unit. Lt. Georg Schlenker, Staffelführer of Jagdstaffel 41 had made the same experience. Therefore, he and Walter Böning coordinated the flight schedules of their Staffeln without actually informing the Kommandeur der Flieger der Armee-Abteilung B, who himself did not seem to mind. In addition, Walter Böning followed the example set by Georg Schlenker in the procurement of equipment for his Staffel by sending his Offizier z. b. V. on "missions of military importance" directly to the aircraft factories, which were happy to support the front-line Staffeln. Spare parts in particular were considerably easier to obtain this way rather than via the official channels.[23]

Right: Lt. Hans Böhning, 3rd from left, with members of the ground personnel at Habsheim.

Four days later, on December 1, 1917, the Staffel succeeded in achieving its first aerial victory. It was cloudy and hazy when Walter Böning and Hans Böhning departed on a flight to the front in response to reports indicating lively enemy flight activities there. They first passed overhead Mülheim before following the railroad tracks leading north-westbound towards Alt-Thann, where they turned southbound into the direction of Niederaspach, looking for enemy aircraft along the frontlines.

This route of flight was routinely followed by Jagdstaffel 76 pilots. As the prevailing winds were northerly, the aircraft initially flew into the wind on the German side, in order to gain altitude. Then, they would turn southbound to fly along the frontlines, supported by the tailwind, to patrol the area. If they spotted enemy aircraft, they were able to make use of these tailwinds to their best advantage in order to attack in a high speed in a dive.[24]

Near Altkirch, they spotted a French Sopwith two-seater flying below them.

Hans Böhning attacked the enemy plane in a dive, while Walter Böning turned to the west in order to cut off the enemy's escape route. Hans Böhning opened fire from both machine guns, which was met with return fire from the Sopwith's observer. Hans

Above: Uffz. Siegfried Walther and the dog of the Staffelführer in the cockpit of their red Albatros D Va with Walther´s personal white "W". This photo allows a closer look at the telescopic gunsight and rear-view mirror arrangements also seen on Karolus Bärenfänger´s and Walter Böning´s Albatros fighters. (R. Kastner)

Böhning skillfully maneuvered himself beneath the two-seaters tail; following a well-aimed burst from his guns, the enemy dropped to the side and entered a steep dive. Hans Böhning followed his opponent, but he was then obliged to turn away due to the fire of French anti-aircraft batteries.[25]

The Sopwith two-seater, which apparently flew with Escadrille Sop. 60, came down in a crash-landing on its own soil, the aircrew remained unharmed. The victory was confirmed as the Staffel's first, as well as being Lt. Hans Böhning's 5th personal aerial victory.

During the following week, rain, snow, and blowing snow hampered flying activities, sometimes preventing flights altogether. On December 2, 1917 Jagdstaffel 77, another newly formed Staffel under the Command of Lt. Walter Ewers, joined the other units stationed at Habsheim.

On December 7, at 15:20, Uffz. Siegfried Walther departed along with another pilot. The weather was partly cloudy and hazy; they followed the railroad tracks toward Sennheim, as usual, in order to patrol along the frontlines. Contrary to reports, no enemy aircraft were to be spotted. Both Albatros then turned west, entering French airspace, and overflew Thann. South of Gewenheim they sighted a large twin engine French Letord with a three-man crew. The aircraft was used for reconnaissance and bombing missions; one machine-gunner was seated in the nose and another in the rear fuselage, behind the pilot. The aircraft therefore possessed considerable firepower, but it was unwieldy and slow. The duo attacked the Letord from below and from the side in order to avoid coming into the field of fire covered by the gunners. The Letord began to curve as both onboard gunners fiercely set about to defend themselves. One of the Albatros turned away in order to avoid the defensive fire, while a salvo fired by Uffz. Siegfried Walther scored a bull´s eye on the French aircraft. His shots must have hit

Above: Another photo of Vzfw. Walther taken a few seconds after the one on the previous page. (Koller)

a fuel tank; moments later, the Letord was engulfed in flames, and plummeted from the sky with a swooshing trail.

The French report covering this incident reads:

"Un appareil Letord de 11 Escadrille A.R. 50 tombe en flammes dans nos lignes - la suite d'un com- bat contre 2 avions ennemis. Le brigadier Machetav, pilote, le sous-lieutenant Moullaard, observateur, et Ie soldat Michel, mitrailleur, sont tus."[26]

"A Letord aircraft of Escadrille A.R. 50 crashed burning into our side of the lines as a result of aerial combat against two enemy aircraft. Crew members were Brigadier Machetav, pilot, 2/Lt. Moullaard, observer, and Soldat Michel, gunner."

Uffz. Siegfried Walther had obtained his first aerial victory, and it was the second one for Jagdstaffel 76.

Two days later, three Jagdstaffel 76 Albatros led by Walter Böning flew across French lines in sunny weather. They were tasked with protecting Rumpler C IV C.1458/17 (Lt. Dagobert Jäger and Oblt. Wilhelm Leonhard), a German reconnaissance aircraft of Fliegerabteilung (A) 289, which was ordered to take aerial photographs of French supply lines along the railway between Oberbruch-Gewenheim. They guarded the German twin-seater by accompanying it using two aircraft flying approximately 100 meters offset and above, while Lt. Böning protected all of the aircraft from a higher altitude trailing slightly behind them. They were flying in this formation when they entered French airspace. Between Thann and Masmünster, they were intercepted by several Spad single-seater aircraft of Escadrilles Spa.49 and Spa.151; a vicious aerial battle ensued. In the course of this, an Albatros was caught by a Spad from behind and pushed lower. Walter Böning, who was about to attack another Spad, recognized the dangerous situation his comrade found himself in, disengaged from his opponent, and proceeded to attack the other Spad that was harassing his comrade. He managed to maneuver into firing position directly behind it

Above: No winter tires fitted! Pilots with the Staffel vehicle just after a fresh snowfall in the winter of 1917–18. Seated from the left: unknown, Uffz. Wagner, Lt. Böning, Lt. Vollert, Vzfw. Hüttner. Note the Bavarian coat of arms carefully applied to the rear door.

and pulled the trigger. Shots hit the enemy aircraft's engine which immediately began to emit smoke; it entered a downward spin. At the same time, Walter Böning was attacked by yet another Spad from behind; his aircraft received multiple hits. He tipped over and entered a nosedive in an attempt to avoid further fire. Suddenly, the shots stopped, and his enemy had disappeared. He turned and watched the Spad fly westbound at a much lower altitude. A light blue Albatros with a black and yellow fuselage band appeared next to him. It was Lt. Wilhelm Kolb, who had rushed to assist his Staffelführer. His intervention had caused the Spad to let go of Walter Böning, and quickly rush for safety by entering a nosedive. Meanwhile, however, the remaining French fighters had attacked the German two-seater and had caused it to crash between Gewenheim and Masseveaux. Its crew died as a result of the combat.

Casualty report:

Lt. Dagobert Jäger (pilot) and Oberleutnant Wilhelm Leonhard (observer) departed on a frontline mission on Rumpler C.IV 1458/17 at 1035. According to the report of an escort aircraft flown by Lt. Heidenreich (pilot) / Lt. Pflügel (observer), the aircraft flown by Jäger / Leonhard came to their aid during an aerial battle above Belfort. Their aircraft was beset by another Spad. It was later spotted approximately 300 meters below. Heidenreich / Pflügel had lost sight of the aircraft during a turn.

According to a report by the Red Cross, the missing crew fatally crashed, and was buried at Michelbach military cemetery (Upper Alsace).[27]

According to French files, M.d.L. Jean Bouyer (Spa.49) and S/Lnt. Henri Hirsch (Spa.151) were credited with the joint kill of a Rumpler.

The Spad which had been hit by Walter Böning and spun down trailing smoke must have managed to make an emergency landing. It appears as though the pilot remained unharmed. For this reason, there are no aircraft recorded as having been lost in the

Above: New Year's Day 1918 party at the Casino of Jagdstaffel 76 with Christmas decorations still in place. Standing, from left; Uffz. Wagner, ?, Lt. Wenig, Vzfw. Hopf. Seated, from left; Lt. Böning, Lt. Böhning, paymaster Schweppech, Vzfw. Bärenfänger.

available French files. However, as a Spad crashing on the French side had been observed from the German side of the front, Böning was credited his 7th aerial victory. After the passage of so many decades, it is no longer possible to verify whether the German observation post actually witnessed a crashing Spad, or if they may have confused a German two-seater shot down near Gewenheim with the crashing Spad.

After landing at Habsheim, Walter Böning counted multiple hits to his Albatros D Va. That evening, he thanked Wilhelm Kolb for coming to his rescue by gifting him a bottle of good Alsatian wine.

In the following days and weeks there were repeated aerial battles, which largely took place in French airspace due to the more defensive fighting tactics of the French. Walter Böning described the situation in a letter to his parents:

"We have had a good number of successes in the Staffel thus far; but we have to look for aerial battles far beyond the front lines, and as a result we are rarely able to rely on observations from our posts on the ground. Nevertheless, a few successful battles have been observed and I am hoping to be credited with two aerial victories."[28]

On December 12, 1917, Walter Böning departed on a flight to the front in French airspace along with four Staffel aircraft at 15:25. At approximately 16:30, they began their descent toward Habsheim airfield following an uneventful flight. As usual, Walter Böning was the last aircraft in the formation and flew at the highest altitude in an effort to protect his aircraft against possible attacks. The Staffel was in the final stage of the landing approach at an altitude of approximately 1000 meters when Walter Böning watched Lt. Kolb's light blue aircraft enter a climb before tipping over to the right and disappearing

New Year's Day 1918 celebration with Siegfried Walther and Walter Böning. Apparently, there was no shortage of Christmas tree decorations in the Fliegertruppe.

behind a hill in a shallow glide. Walter Böning assumed that Wilhelm Kolb had made an emergency landing due to technical problems. He landed, allowed his aircraft to roll out before jumping out of the plane and getting into one of the cars parked next to the hangars. He drove to the spot where he had watched Kolb's aircraft disappear behind the hill. When he arrived, he was shocked to discover Lt. Kolb's Albatros had a completely destroyed nose area. German soldiers who had rushed to his rescue were in the process of recovering the pilot from the wreckage. A field medic who examined the seemingly unconscious pilot, shook his head. Lt. Wilhelm Kolb had been killed on impact due to skull fracture.

Following is the Jagdstaffel 76 accident report:

"Lt. Kolb departed on Albatros D.V 5294 (the aircraft was a D Va, author's note) at 3:25 in the afternoon along with several Staffel aircraft. At 4:30, all of the aircraft returned from their fighter patrol flight. At an altitude of approximately 1000 meters, it appeared as though Lt. Kolb was attempting to execute a loop. Observers on the ground reported that he pulled back hard but began pushing the aircraft back down immediately after. It was further observed from the ground that the tip of his right upper wing had broken, pointing upwards. Lt. Kolb's aircraft entered a shallow glide in the direction of Habsheim in order to land. The impact of the aircraft could not be observed. It is assumed that Lt. Kolb was unable to recover his aircraft from the bank which resulted from the broken wing as the control cables must have been jammed. It is likely that the aircraft's left wing impacted the ground which snapped off the landing gear. The aircraft flipped over, causing the engine to penetrate the cockpit, which in turn caused the serious fractures of the legs. Lt. Kolb was not wearing a helmet; his head hit the frozen ground, causing his skull to fracture."[29]

That evening, no concert was being held at the Jagdstaffel 76 casino.

Wilhelm Kolb's body was transferred to his hometown of Munich, where he was buried with military honors in the honorary section of the Waldfriedhof cemetery.

Deeply saddened, Walter Böning wrote about the accidet on December 17, 1917:

"The sad news of Adolf Tebbe's death (a school friend of Walter Böning) coincided with the fatal crash of my best and dearest pilot, Lt. Kolb. Wing failure occurred on his aircraft and it crashed, resulting in his death. I would have rather he had fallen in aerial combat..."[30]

In conversation with Bruno Schmäling, he remarked regarding this particular crash that he was not putting blame on anybody, but the question remained whether better trained mechanics would have recognized the possible damage to the wings during a routine inspection. In addition, the delivered spare parts were deficient or of poor quality. Particularly some of the metal parts had been made of a softer metal which was prone to bend under stress.[31]

On a final note, Unteroffizier Karl Hopf transferred to the Staffel on the same day. He was from Würzburg in Unterfranken where he was born on July 13, 1894. Before the war, he had joined the Royal Bavarian 9th Infanterie-Regiment on October 22, 1913 with which he deployed when the war began. He volunteered to join the Fliegertruppe and was transferred to undergo pilot training with Bavarian Flieger-Ersatz-Abteilung 1 in Schleißheim on October 26, 1916. He attended additional training with the Militär-Fliegerschule in Lachen-Speyerdorf in the Bavarian Palatinate. On June 18, 1917 he was first transferred to Armee-Flugpark C, followed by a transfer to Armee-Flugpark 4 on July 13, 1917, and lastly to Bavarian Schutzstaffel 26 on July 17, 1917. As he had signed up to join one of the fighter units, he was sent to attend Jagdstaffelschule I via Armee-Flugpark 4 just one month later. From there, his journey led him to Jagdstaffel 76.[32]

In return, the Staffel had to part with Lt. Franz-Xaver Danhuber on December 13, 1917. He was transferred to Fliegerschule I in Schleißheim, where he was given the task of forming Royal Bavarian Jagdstaffel 79 on January 7, 1918; he would experience the same bureaucratic troubles with the Adminstration of Bavarian Flieger-Ersatz-Abteilung in Schleißheim that Walter Böning was obliged to deal with when establishing his Staffel.

Four days later, orders reached the Staffel that Erwin Wenig (who had been promoted to Oberleutnant in the meantime) had been designated Staffelführer of newly-formed Royal Bavarian Jagdstaffel 80. The new Staffel was being assembled at Bavarian Flieger-Ersatz-Abteilung 2 in Fürth. The orders indicated that Wenig was to proceed to Fürth immediately in order to accelerate the process of Staffel formation. As mentioned by Max Speidel, former Offizier z. b. V. of Jagdstaffel 80, in conversation with one of the authors, the formation of the Staffel in Fürth was a relatively smooth process, once initial difficulties had been overcome.[33]

As a result of this double transfer, the Staffel had lost two experienced pilots which had been personally requested by Walter Böning; by the turn of the year, the Staffel had six pilots at its disposal, which was less than half the nominal strength of a Jagdstaffel. Initially, this did not hamper the ability of the Staffel to carry out its missions. During the following days, the skies darkened, and winter arrived, bringing violent snowstorms. In a letter he wrote home on January 2, 1918 Walter Böning mentioned that flying during this poor weather was practically impossible:

"It is winter here, thick blankets of snow covering everything, low clouds, mild frost-no flying conditions, but there is enough work to be done otherwise. We have not flown in two weeks, with the exception of December 24; therefore, there are no further aerial victories to report. So far, we have achieved three aerial victories, including one for me, a Spad which I shot down in flames, 14 kilometers behind enemy lines. A Nieuport, which was even further away, but unfortunately it had not been observed or photographed by an accompanying reconnaissance aircraft as I had shot it down. Confirmation for this one is still pending, but I do not believe that I will receive credit for it."[34]

His assessment of the situation would turn out to be correct, he was not credited with the kill of the Nieuport (the exact date and detailed circumstances regarding this kill have not been passed on, unfortunately), due to a lack of witnesses.

Until January 11, 1918 the Staffel was only able to fly on January 3, 4, and 6, 1918:

"For several days, we were almost entirely cut off from the rest of the world. The telephone was the only way for us to have a conversation with fellow human beings also buried under heaps of snow. A blanket of snow at least 50 cm thick rendered all car usage impossible, and every walk a torture. It was an uncomfortable time; people wanted to live but the Provisions Department is 9 km away which means that all the bread, meats, etc. had to be carried here by the people. Truly, this was no joy. Of course, we were not able to fly in spite of fair skies at times. The aircraft would get stuck on takeoff roll or turn over on landing. The French were in the same boat, as we did not see any of them up in the air. The temperatures are rising, and yesterday, the snow began to melt. I am glad that our barracks are elevated about 50 cm above the ground, otherwise we would constantly be taking footbaths. Everything is flooded, our airfield has turned into a lake."[35]

This quiet time allowed pilots some time to rest; each night, music was played at the Casino barracks. Oftentimes, pilots and observers from other Staffeln, along with female auxiliary staff and nurses, were invited to join as guests, if they were willing to take on the walk through the deep snow. Clemens Vollert tried everything in his power to provide abundant food and good wine to everyone.

For the Staffelführer, the deep snow could have meant some downtime, if he did not have continued troubles with his superiors in Armee-Abteilung B:

*"Everything has largely remained the same here, lots of work and even more trouble. Lately, I have encountered some challenges with a partly superior Hafenkommandant (*Kommandeur of Habsheim airfield, author's note*). As all Staffeln on the airfield voice identical complaints about him, we will try to put a stop to this lout's game. He is the oldest Flieger-Hauptmann of the Army, older than the Kommandeur (Kommander der Flieger of Armee-Abteilung B, author's note); therefore, he does not like being the Abteilungsführer (of a double-seater unit) of 6 aircraft while younger Leutnants have 14 aircraft at their disposal. If I ever had a superior like him when the war is over (he is Bavarian), I would immediately quit my active duty service. I have largely given up on such kinds of plans for the future as it is."*[35]

In conversation with Bruno Schmäling he added:

"After everything I had witnessed in Schleißheim and Habsheim, becoming a career officer was no longer an option for me. Studying medicine and becoming a doctor increasingly emerged as my plans for the future."[36]

During the following weeks, missions were only flown on a handful of days. However, the sorties flown in January 1918 did not yield any viable results.

3 Aerial Combat over the Vosges

Armee-Abteilung B, February 1918 – March 1918

Above: Jagdstaffel 76 members in a festive mood at the Casino. From left to right: Lt. Böhning, Lt. Böning, Vzfw. Wagner (back), unknown, unknown, unknown, Lt. Wenig, paymaster Schweppech and Vzfw. Bärenfänger.

After the initial ground fog had cleared on Februay 2, 1918, a Kette of the Staffel departed under partly cloudy skies, patrolling initially along the frontlines heading north. When they did not encounter any French aircraft there, they turned towards the west, entering French airspace. There, they spotted four A.R.2 two-seaters escorted by Spad single-seat fighters. When the Albatros attempted to attack the A.R.2, they were in turn attacked by the Spad fighters that moved in front of the A.R.2 in order to screen them; a violent aerial battle ensued. Walter Böning observed how a Spad which he had attacked entered a steep dive and appeared to land. He then attacked one of the A.R. two-seaters, but return fire from this plane hit his engine, forcing him to turn away. Upon his arrival at the home field, he reported the following observations:

"We were finally able to do some real flying yesterday. I came back without a single round left in my guns. Only one Spad forced to land, but I received shots through the propeller into the engine by an A.R. two-seater; only a compression lever was torn off as a result. I was at an altitude of 400 m, 8 km beyond enemy lines, and I was greatly relieved when my engine carried me back to the airfield at half of its RPM speed."[1]

The French report read as follows:

"Au cours d'un combat livre par 4 avions de l 'Escadrille A.R. 230 e une patrouille ennemie, un Albatros parait sérieusement touché et piqué dans ses lignes.»

"During an aerial battle between four Escadrille A.R. 230 aircraft and an enemy patrol, one Albatros was badly hit, dropping into a steep nosedive across

Above: Vzfw. Wagner with his ground crew in front of his green Albatros D III (O.A.W.). The tail of another Albatros D III (O.A.W.) can be seen at left. The small gap between the spinner and the forward fuselage of the aircraft shows well in this view.

Above: Vzfw. Paul Wagner wearing his Chevaulegers Uniform. He fatally crashed at Colmar airfield on 4 February 1918.

his lines."[2]

According to information supplied by the French report, the victory report indicated a "likely aerial victory" only.

Two days after the aerial battle with the French aircraft, Vzfw. Paul Wagner departed on a flight toward Colmar. Overhead the Colmar north airfield, he displayed a small air show by rolling the aircraft multiple times and executed loops at low altitudes. During a loop, which he had started at too low of an altitude, he was unable to recover the aircraft in time, resulting in a crash at 10:46. The aircraft immediately went up in flames upon impact. Ground personnel rushing to the scene were not able to prevent the aircraft and its pilot from being gutted.

Walter Böning was upset:

"Unfortunately, I have lost another pilot (in addition to two transfers, author's note), a Vizefeldwebel who sadly has himself to blame for the most part. He frequently flew at low altitudes like a loose cannon. Even though I forbid his antics overhead other airfields, he could not resist the temptation of playing the fool over enemy airfields. Aircraft and pilot completely burned."[3]

On the same day five Staffel Albatros departed south westbound under the command of their Staffelführer. They passed Altkirch in order to patrol along the frontlines until reaching the Swiss border. When they were roughly abeam Hirtzbach, they spotted a squadron of numerous French aircraft flying along the frontlines. The Staffel attacked the French formation, which consisted of 7 Sopwith

Above: Funeral services held in honour of Vzfw. Paul Wagner, with several members of his unit in attendance. (R. Kastner)

two-seaters from Escadrille Sop. 129. Walter Böning succeeded in shooting at one of the Sopwith referred to as Spad two-seater. It caught fire and crashed on the French side, near Largitzen. This aerial victory was Lt. Walter Böning's 8th, and the 4th for the Staffel. As confirmation of application for recognition with the Kogenluft seemed to take more time than anticipated, it was officially confirmed as the 9th aerial victory.[4]

The French documents referring to the incident read as follows:

"34 Sopwith biplaces du Group de Bombardement 4 ont effectué un barrage permanent entre Thann et la Suisse. Deux combats ont été livrés par nos avions: l'un par 9 Sopwith contre 5 biplaces ennemis l'autre, par 7 Sopwith contre 3 Albatros monoplaces. Au cours de ce dernier combat, un Albatros a piqué fortement dans ses lignes, paraissant touche.

Un Sopwith de l'escadrille Sop. 129 (G.B. 4) est tombe en flammes dans nos lignes le sergent pilote Cordonnier est très grièvement blessé, le mitrailleur Allais est tu. "[4]

"34 Sopwith two-seaters of the Bombardment Group 4 executed a permanent barrier between Thann and Switzerland. Two combats were carried out by our aircraft: one involving 9 Sopwith against 5 enemy two-seaters, the other involving 7 Sopwith against 3 single-seaters. During the latter battle, one Albatros crashed into its own lines. One of our Sopwith from Escadrille Sop. 129 (GB 4) went up in flames. The pilot, Sergeant Shoemaker and gunner Allais were severely wounded.[5]

The following notes are supplementing the above report: the badly injured pilot in the Sopwith shot down by Walter Böning later died at the field hospital as a result of his injuries. The Albatros which was mentioned to have been shot down (no losses were reported by Jagdstaffel 76) was only credited as a "probable" to Escadrille Sop. 129.

One week later, on February 11, 1918, the Staffel was heard of again.

Walter Böning described the events in detail in a letter to his parents:

"I had been flying along the front for about 1 ¾ hours along with two other aircraft without an

Above: Lt. Hans Böhning and Lt. Walter Böning enjoy a cigarette in front of the pilot's casino at Habsheim. Note the two bottles of wine leaning against the bottom of the Staffel hut constructed of birch trunks.

enemy *in sight. We flew at an altitude of 4500 meters when I suddenly spotted two aircraft circling west of Altkirch. After a few nosedives and side spins, I managed to get close enough to determine the aircraft type. The two French had sighted me also, and already opened fire at me from a distance of 400 m. They alternated shooting and altitude; finally, I picked off one them, and fired 300 rounds; the aircraft went up in flames. Meanwhile, I had found myself beneath the second enemy aircraft whose observer merrily sprayed me with bullets from a distance of 50 m. Shortly after my opponent was dealt with, I took a hit that pierced the cooling pipe, causing all the water to drive out in a mighty cloud of steam. First, I thought I was on fire but when the hot water burned my face, I was relieved. I was at an altitude of 1200 meters, and 4 kilometers from the frontline. I had begun the battle at 1500 meters. I realized that my engine would soon quit and was concerned about making my way back across the frontlines. Thank God, the engine kept running until I reached the frontline, thanks in part due to strong westerly winds. But soon after I had to get down as my propeller had stopped spinning. I made an emergency landing over hills and valleys; the aircraft did not get away unscathed, however. The landing gear was damaged, and the propeller broke upon collision with a fence post.*

Fortunately, I was unharmed, and I immediately began looking for my other two aircraft. People rushing to my aid told me that there were four of them coming down in total, including mine. The best-case scenario was that the second enemy aircraft had been shot down as well, after it had shot down one of our Albatros, or forced it to land. The people said that one of the last aircraft had landed just shy of our lines. And, oh no-they were right.

Uffz. Walther who had attacked beside me had focused his efforts on the second aircraft whose observer was busy firing at me; he shot him down in flames as well. However, as his interrupter gear shaft, which permits for precise MG shooting through the propeller had broken off during the combat, he ended up completely shooting off his own propeller, which forced him to go down as well. He had no engine power left, and entered a glide towards the East, pushed by the tailwind. He managed to land just behind the French barbed wire fences, right between the German and French lines but ended up turning over and flipping his aircraft onto its back. He then ran, as fast as he could, across the shot-up area, from one shell crater to the next, until he safely reached the German trenches.

Both of our damaged aircraft were salvaged the following night; they are already back in my hangar, of course in need of lengthy repairs."[6]

The French report on the aerial battle reads as follows:

"Au cours d'un combat entre 3 Albatros et 2 avions A.R. de l'escadrille 230 (pilotes, maréchal des logis Monnier et sous-lieutenant Goeury; observateurs, sous- lieutenant Taront et sous-lieutenant Berger) un des avions ennemis est abattu entre les lignes et un autre probablement abattu dans ses lignes.

Au cours du combat, le maréchal des logis Monnier et Ie sous-lieutenant Tartont sont blesses, le dernier grièvement. "[7]

"During aerial battle between 3 Albatros and 2 aircraft of Escadrille A.R 230 Squadron (pilots, Marschall Monnier and Oberleutnant Goeury, observers Unterleutnant Taront and Oberleutnant Berger), one of the enemy aircraft was shot down

Above: Hans Böhning strikes a pose for the benefit of the camera in front of Uffz. Josef Diem's black and yellow ringed Albatros D Va in February 1918. The hangars at Habsheim are barely visible in the distance.

Above: Signed photo of Vfzw. Josef Diem. According to Walter Böning, he was one of the best and most reliable pilots of the Staffel. (R. Kastner)

between the lines, and another over its own territory. Sergeant Monnier and Unterleutnant Tartont were injured, the latter gravely."[7]

Lt. Walter Böning was credited with the kill of a French "Paul Schmitt" at 16:50 in French territory near Füllern; it was his 8th aerial victory. Uffz. Siegfried Walther was credited with bringing down of the second "Paul Schmitt" on French territory near Carsbach; this was his 2nd aerial victory.

The French credited two crews with the kill of an Albatros, while the second one was only reported as "likely shot down".

Walter Böning had already penned two proposals for the bestowal of the EK I to Uffz. Siegfried Walther; on both occasions, he was denied the award, in spite of the fact that, according to Walter Böning, "he has already shot down an aircraft, and he is an impeccable soldier." Walter Böning was angered excessively by this, which only intensified his resentment of the Armee-Abteilung B Staff. Finally, following a third application, and a personal visit paid by Lt. Böning to the Kommandeur der Flieger, Uffz. Siegfried Walther was awarded the EK I on February 12, 1918. Four days later, he also received the honorary goblet "Der Sieger im Luftkampf" (victorious in air battle).

On February 13, 1917 the Staffel received personnel reinforcement in the form of the arrival of Unteroffizier Josef Diem. Josef Diem was born in Munich on November 27, 1893. In 1912, he joined the Bavarian 2nd Telegraphen-Bataillon, moving to war in 1914. In the spring of 1917, he joined the Fliegertruppe; on August 4, 1917 he was transferred to Schleißheim for pilot training with the Bavarian Flieger-Ersatz-Abteilung. He spent the initial part of his pilot training at Bavarian Militär-Fliegerschule 1, followed by a stint at Fliegerschule 2 in Lachen-Speyerdorf, where he completed his training. Following his transfer to AFP B on February 2, 1918 and the then-usual additional 10 day fighter pilot training course at Jagdstaffelschule I, he transferred to Jagdstaffel 76 on February 13.[8]

Walter Böning recalled him as being a very talented and spirited pilot who was highly reliable during missions. While he was a rather inexperienced flyer, he soon developed into one of the main pillars of the Staffel.[9]

On February 16, 1918 Leutnant der Reserve Hans Hertrich arrived at the Staffel. He was from Nuremberg in Central Franconia, born on April 15, 1891. At the outbreak of the war, he was an infantryman with the Bavarian 14th Infanterie-Regiment, working his way up to Officers ranks within just over a year of service. Following his application to join the Fliegertruppe, his transfer to Flieger-Ersatz-Abteilung in Schleißheim was ordered on May 31, 1917, where he successfully completed pilot training with Fliegerschule 1, followed by a transfer to Aflup. B, and additional single-seater training at Jagdstaffelschule I in Valenciennes.[10]

The same day, Clemens Vollert was unexpectedly transferred to Royal Bavarian Jagdstaffel 78, where he was to serve as Special Duties Officer. He returned to Jagdstaffel 76 just nine days later.

On February 17, Lt. Hans Böhning attacked an A.R.2 near Dammerkirchen over enemy ground, which went down as a result of his accurate gunfire. The victory was confirmed to him as "forced to land behind enemy lines".

Furthermore, on February 17, the newly established Royal Prussian Jagdstaffel 71, commanded by Lt. Hermann Stutz, arrived in the Army sector and settled down at Colmar airfield.

As much as personnel reinforcements were welcome, these also meant that the remaining experienced pilots, such as Karolus Bärenfänger and

Above: Walter Böning and Hans Böhning (first and second from right) enjoy a sunny and warm winter's day in Habsheim with some comrades.

Below: Staffel pilots celebrating the promotion of Karolus Bärenfänger to Leutnant der Reserve. Seated, from left: Lt. Bärenfänger, paymaster Schweppech, unknown, Lt. Böhning, Vzfw. Walther, unknown. Standing, from left: unknown, Vzfw. Hüttner and Uffz. Hopf.

Above: Leisure time at Habsheim; front left, Lt. Hans Böhnung, Lt. Helmut Dietlen, Jasta 41, 3rd from left.

Siegfried Walther, now had to keep a close eye on the newbies, which somewhat weakened the Staffel's fighting strength. This was especially true after Lt. Hans Böhning's transfer to join Jagdstaffel 79 as Staffelführer, on February 23, 1918. The previous Staffelführer of Jagdstaffel 79, Leutnant Franz-Xaver Danhuber, had been badly injured in a crash at the airfield on February 11, 1918. While Walter Böning was happy that his friend was entrusted with leading his own Staffel, he was also deeply saddened by his departure. His leadership skills and his technical knowledge would have been very helpful assets to Jagdstaffel 76.

On the same day, 21-year-old Gefr. Xaver Hüttner was transferred to the Staffel. He was from Wörth on the Danube River. He had joined the infantry as a volunteer when the war began. Following his successful application to join the Fliegertruppe, his career path first led him, like many others, to Schleißheim in May of 1917, followed by a transfer to Militär-Fliegerschule 4 in Lager Lechfeld, where he successfully completed pilot training. Following his transfer to AFP B in February 1918, and some finishing training touches to his skills at Jagdstaffelschule II, Hüttner joined Jagdstaffel 76 on February 23, 1918.[11]

On February 26, a Kette led by the Staffelführer crossed the front in good weather in order to search for enemy aircraft over French territory. All of a sudden, Uffz. Josef Diem departed the formation and attacked a French aircraft he had sighted. Walter Böning spotted the A.R. just moments later, following Diem along with a third pilot. Josef Diem was the first to reach the enemy aircraft and opened fire. The A.R. went down "stinking", meaning that it was leaving a trail of thick smoke, obliged to make an emergency landing on French territory. Josef Diem followed the aircraft on its way down, and only gave up the chase after French anti-aircraft artillery opened fire on him. Walter Böning then reassembled his Kette in order to continue the patrol flight along the front, before finally returning to Habsheim.

Of course, the Staffelführer was proud of Josef Diem's success as he had proven himself to be instilled with fighting spirit right from the beginning

Above: Six members of the Staffel on a walk at the edge of Habsheim airfield. Lt. Clemens Vollert can be seen at left, holding his camera into the camera. Lt. Hans Böhning, 3rd from left, Vzfw. Walther, 4th from left, and Lt. Karolus Bärenfänger, at far right.

of his career as a fighter pilot. Nevertheless, he pulled him aside after landing and insistently cautioned him to curb his recklessness just a bit. Prior to starting an attack on a lone flying double-seater, it was imperative to ascertain that it was not escorted by fighter aircraft flying at higher altitudes, which was often the case with the French Escadrilles. Ignoring this fact would easily result in falling prey to Nieuport or Spad single-seaters jumping down from above. According to an available French report, Josef Diem's opponent could have been an aircraft from Escadrille Sal. 1, whose observer, Asp. Murot, was wounded in aerial combat. To his misfortune, the victory was claimed at too great a distance behind enemy lines, and no observers from the ground witnessed the event. Therefore, he was not allowed official confirmation for this victory claim.

In a letter from March 7, 1918 Walter Böning described that the German side, too, experienced losses in aerial combat:

"Two days ago, the French were successful here for the first time in a long time; they forced two of our aircraft to land, resulting in the deaths of both observers, and grave injuries of both pilots. Thereupon, I had received orders yesterday to utilize the Jagdstaffeln for three hours during our prime duty hours for Ifl. (Infanterieflieger, Infantry

Above: Jagdstaffel 76 at Habsheim airfield in March 1918, pictured in front of Lt. Walter Böning's "baby blue" Albatros D Va. From left: Vzfw. Hüttner, Vzfw. Walther, Lt. Bärenfänger, Lt. Böning, Lt. Vollert, Lt Hertrich, Uffz. Diem, paymaster Schweppech, Vzfw. Hopf. Böning´s reserve Albatros D III (O.A.W.) can be seen at left.

Above: Five Albatros D Va and D III (O.A.W.) provide the backdrop for the Jagdstaffel 76 members, photographed within moments of the previous picture. Seen from left: Vzfw. Diem, Lt. Bärenfänger, Vzfw. Hüttner, Lt. Böning, Lt. Hertrich, Lt. Vollert (behind), paymaster Schweppech, Vzfw. Hopf. Above: Vzfw. Walther. The aircraft closest to the is the Albatros D Va of the Staffelführer.

Right: A very graphic image capturing the wreckage of the Escadrille Sop. 252 Sopwith two-seater shot down by Lt. Walter Böning near Lauterbach on 15 March 1918. The charred remains of the fuselage and one of the unlucky crew members got stuck in some trees and were soon covered by some of the last snow to fall this winter.

co-operation aircraft) and Afl. (Artillerieflieger, artillery-supporting aircraft). Currently, I am the acting group leader of all Jagdstaffeln present here. Well, during that time, five enemy aircraft were shot down, usually in flames."[12]

After the weather had improved on March 6, 1918 Walter Böning departed on a front flight at 13:00 along with five of his men. He later shared the following account of his eventful flight in a letter to his parents:

"I departed at approximately 13:00 with two Ketten. Unfortunately, several of my young pilots disappeared after three quarters of an hour, as they either lost sight of me, or had experienced engine trouble. In the end, one Uffz. and myself were the only aircraft left of my Kette at 13:45. Incidentially, for the time being, Uffz. Diem is my best pilot; he is spirited but unfortunately he is still a somewhat inexperienced Kettenflieger.

I flew towards the north and was approximately 3 km from the mountains when I spotted three Spad dive down upon me. They opened fire at a distance of 400m even though we were flying toward each other. I immediately throttled down, which led me to hang in the air like a ripe plum. I expected the Spad (suddenly, I was alone, as Diem slipped off in a turn) to attack from behind, and at the same altitude; which is exactly what happened. But, as soon as they had turned toward me, I opened full throttle, and thanks to the sudden surplus power, out-climbed the three by 50 meters in no time in a left-hand turn. I picked off one of them and opened fire. He entered a spin after only 30 rounds. (So far, I have not been able to determine whether I shot him down). Then, I took on the second one; we began a dog fight. I only fired when I had him in my sights. All of a sudden, all I saw were the wings of my opponent - the fuselage, along with the pilot, had decided to go its own way and whooshed away.

Next up, he (the third Spad) pulled up close, and indeed he managed to shoot a single little bullet through my machine, right behind my back. I attacked him in a tight right-hand turn, and as a consequence the Spad entered a right-hand turn as well. This was a somewhat stupid idea, as Spads are not very good at right-hand turns. After a short while, he got too uncomfortable turning to the right; he probably tried to transition into a left turn, and for a short moment he flew straight ahead. He should not have done that, for shortly after, he had a good load shot into his crate, and with a brightly burning trail resembling a comet, he fell into French battery positions near Rodern."[13]

The French "Compte Rendu" reported concerning the events:

"Deux de nos avions sont tombés dans nos lignes à la suite de combats. Les pilotes, sous-lieutenant Dorizon et caporal Brun, ont été tués.»

"Two of our aircraft crashed in our lines after being involved in aerial combat. The pilots, Second Lieutenant Dorizon and Corporal Brun were killed."[14]

S/Lt. Henri Francois Dorizon was a member of Escadrille Spa. 150, while Brig. Paul Brun flew with the previously mentioned Escadrille Spa. 151, both Escadrilles were component units of Group de Combat 16 at the time. The three Spad aircraft were a mixed formation; it is possible that a new formation leader from one unit was intended to gain experience by flying at the front under the supervision of two experienced pilots from a different squadron. This was overall a reasonable and sensible procedure. In this particular case, however,

Above: Following Lt. Hans Böhning's departure from Jagdstaffel 76 in February 1918, his Albatros D Va was taken over by Uffz. Karl Hopf. The "Bavarian Blue" paint used to apply the tail unit marking was very glossy, a reflection of the rudder can be seen clearly in the original print of the photo. Here the aircraft is set up for machine gun alignment.

Above: Detail picture of the Albatros tail mounted to the trestle. The careful application of the Bavarian lozenges can be seen clearly.

they had the misfortune of encountering a highly experienced opponent. The SPADs were confirmed as Lt. Böning's 10th and 11th aerial victories.

Finally, around March 10, 1918, Jagdstaffel 75, which had been newly formed and which was commanded by Oblt. Hasso von Wedel, arrived in Habsheim.[15]

As Habsheim airfield had been quite battered by frequent flying activities, repairs had to be carried out; as a result, Jagdstaffeln 41 and 76 were transferred. It is possible that only the flying units were relocated, for example to Colmar-South. Furthermore, Uffz. Siegfried Walther was promoted to Vizefeldwebel on March 8, 1918.

When the Staffel departed for a front flight under sunny skies on March 13, Walter Böning was concerned, as he always was on this date, even though he was a down-to-earth and logical thinking person:

"Today is the 13th, and even though I am not too superstitious I could still recount many experiences that happened on the 13th day of various months in

Above: Uffz. Karl Hopf and his ground crew with Hans Böhning's former Albatros D Va. After he took over the aircraft, Uffz. Karl Hopf had his personal insignia, the coat of arms of Hochstift (central administration of the archdiocese of) Würzburg, painted onto the fuselage sides.

Above: Uffz. Karl Hopf in the cockpit of his Albatros D Va. Clearly visible is the coat of arms of the "Hochstift Würzburg".

Above: Uffz. Karl Hopf's first airplane mechanic in front of the colorful Albatros D Va. Seen behind it is the white and blue Albatros D Va flown by Lt. Bärenfänger, followed by the two fighters flown by the Staffelführer. (G. VanWyngarden)

the past. When I was part of Jasta 19, I was forced to land on the 13th flight with my machine on different occasions. Today I almost experienced a similar thing. I was flying to the front with Uffz. Diem and decided to anger seven Spad which flew about 300 to 400 meters above us. None of them attacked even though we were already 3 km behind their lines. Finally, one of them swallowed the bait, and I may have hit him a few times as he entered a spin and descended, followed by Uffz. Diem. Now I had to keep six Spad at bay which was not too much of a challenge, as they did not dare follow Diem because I started shooting as soon as they did. Diem finished off his opponent, and finally will get credit for his long sought after initial success. Now, he was at 800 m above the enemy airfield. We returned home while under heavy anti-aircraft fire. I am fine. I almost forgot to mention the Bowden cables on my MG tore which kept me from shooting at the end. If the Spad had known that, my day this 13th of the month might not have been as much fun as it was."[16]

The Spad kill was not reported in the copy of the Staffel war diary nor can it be found in any other German documents. This means that, again, Uffz. Josef Diem did not receive confirmation for this victory due to a lack of German ground witnesses reporting the crash of his opponent.[16]

Two days later Walter Böning took off in cloudy weather with Uffz. Karl Hopf as well as one other pilot; they flew along the railway line from Mulhouse to Sennheim where they turned to the north-west toward Gebweiler in order to fly along the frontlines to the south. When they were north-west of Gebweiler and still over German territory, they spotted two Sopwith two-seaters, escorted by a Spad fighter. The German fighter aircraft flew at a higher altitude and jumped their opponents from above. Walter Böning attacked the rear Sopwith. He watched as several of his rounds hit the aircraft. It dropped on its side, caught fire and then fell ablaze. The second Sopwith two-seater had turned toward him, its observer firing at Walter Böning´s Albatros. Böning dropped below his opponent and returned fire on the French aircraft, which in turn dropped to its side and crashed, flipping over multiple times on its way down. Karl Hopf had attacked the Spad escort, their aerial battle shifting south-west towards Alt-Thann where the Spad spun down after apparently

A Jagdstaffel 76 mechanic in front of an Albatros D III (O.A.W.) at Habsheim airfield.

Above: Jagdstaffel 76 Albatros D III (O.A.W.) and D Va in March of 1918 at Habsheim airfield. The first aircraft is the black Albatros D III (O.A.W.) formerly flown by Oblt. Wenig. Next to it, the red Albatros D Va with Uffz. Walther's white "W" insignia, followed by the white and blue Albatros D Va flown by Lt. Bärenfänger, the yellow and black Albatros D Va flown by Uffz. Diem, and lastly, the two aircraft flown by Staffelführer Lt. Walter Böning.

Above: Enlargement of the red Albatros D Va of Vzfw. Walther

Above: Enlargement of the white and blue striped Albatros D Va flown by Lt. Bärenfänger.

being hit.

Walter Böning briefly wrote the following note on the occurrence:

"Yesterday, three of us attacked a French squadron of three aircraft, shooting down all three of them, two of them by myself. One of them is lying on our side, completely charred."[17]

The French side reported the aerial duel as follows:

"Le sergent pilote Gacon et Ie lieutenant observateur Malot sont tombes dans les lignes ennemies ~ la suite d'un combat soutenu par trois de nos appareils contre sept avions ennemis."

"Pilot Sergeant Gacon and Lieutenant-Observer Malot crashed across enemy lines after three of our aircraft were involved in aerial combat with seven enemy aircraft."[18]

The crew mentioned above most likely came from Escadrille Sop. 60. The second crew, probably also members of Sop. 60, appears to have remained

Above: Side-view of Uffz. Josef Diem's black and yellow ringed Albatros D Va, enlarged from the line-up photo.

Above: Group photo of Jagdstaffel 76 mechanics at Habsheim airfield. The two Feldwebel mechanics who were instrumental in training the Jagdstaffel 76 ground crew at Schleißheim are pictured in the middle of the photo. (R. Zankl)

Above: Mechanics of the Staffel in front of an Albatros D III (O.A.W.) with a dark-painted fuselage. Most likely, this was the aircraft formerly flown by Obltn. Wenig. Who took over his aircraft after his departure of the unit in late January is not known. (J. Leckscheid)

Above: Loading of Jagdstaffel 76 aircraft on March 25, 1918 in Mulhouse for transport to the 2nd German Army. On the right, Uffz. Diem's new Albatros D Va D. 5787/17, marked with the "Münchner Kindl" insignia (the Octoberfest variant of the city of Munich's coat of arms). The tarp-covered prop, still fitted to the white spinner, is stored below the fuselage. On the right, Lt. Bärenfänger's white and blue striped Albatros D Va. Note the box-frame packaging for the Albatros horizontal tail surfaces below the nose of Bärenfänger's aircraft.

unharmed. The Spad which had been reported as shot down by Uffz. Kolb likely entered a spin to escape, and landed on the French side, unharmed.

These were the last aerial victories reported by the Staffel in the area of Armee-Abteilung B, being credited to Walter Böning as his 12th and 13th, and to Karl Hopf as his first. On March 19, 1918, another order was issued to prepare the Staffel for transfer. The mechanics disassembled the aircraft, packed up materials, and transported everything to the railway station at Mulhouse.

Up until this time, Jagdstaffel 76 had been credited with 11 confirmed aerial victories; two pilots had been killed as a result of accidental crashes.

Above: Lt. Bärenfänger and Lt. Vollert at Habsheim.

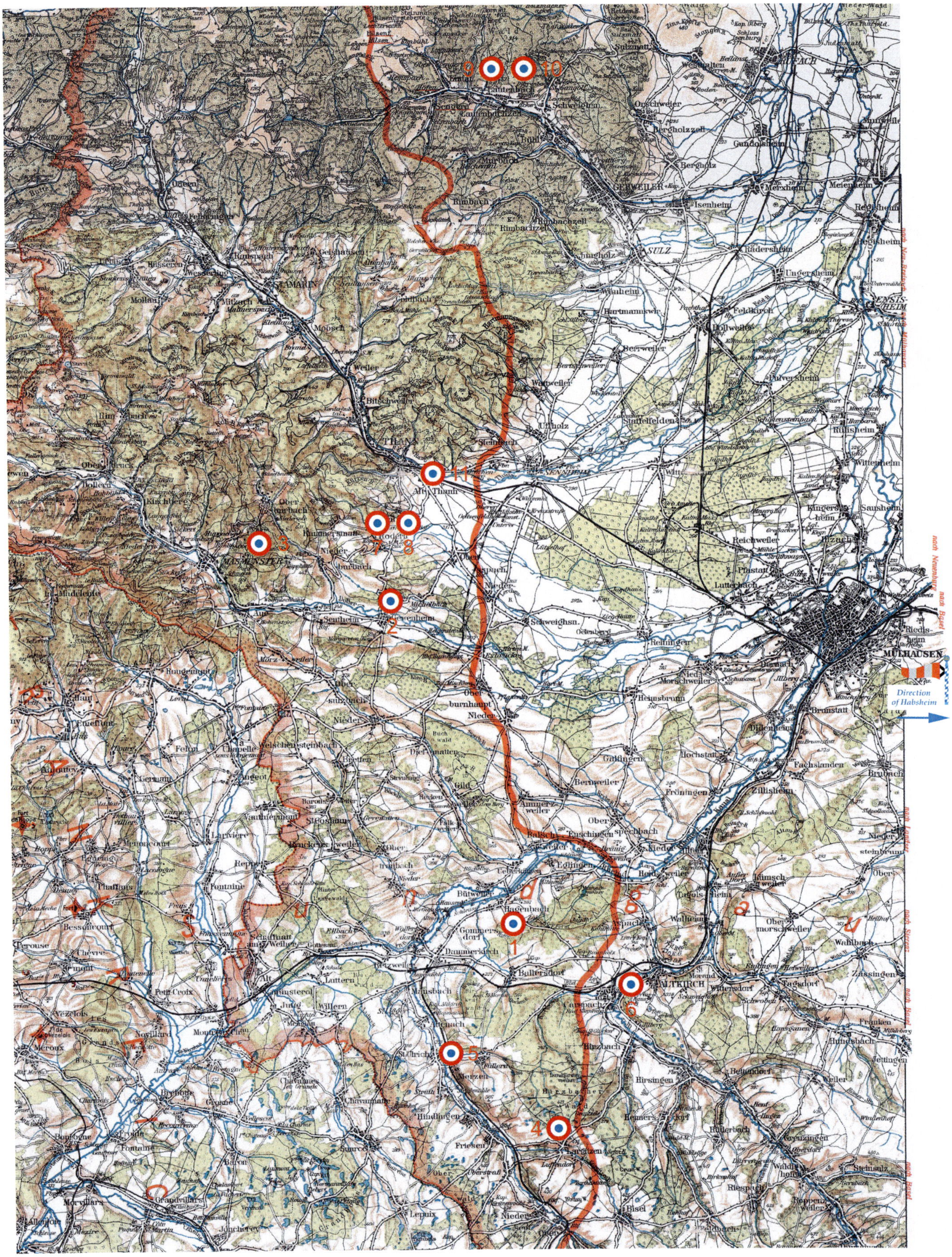
9
10
11
3
7
8
2
1
6
5
4
THANN
MÜLHAUSEN
ALTKIRCH
SULZ
Direction
of Habsheim

Habsheim Airfield
Locations & Events

#		Date	Pilot	Location	Footnote
1.		01.12.17	Lt. Böhning	Hagenbach	Sopwith 1½ Strutter
2.		07.12.17	Uffz. Walther	Gewenheim	Letord three-seater
3.		09.12.17	Lt. Böning	Maasmünster	Spad single-seater
4.		11.02.18	Lt. Böning	Füllern	Dorand AR. 2
5.		11.02.18	Uffz. Walther	Carspach	Dorand AR. 2
6.		04.02.18	Lt. Böning	Largitzen	Spad two-seater
u/c		26.02.18	Uffz. Diem	?	Dorand AR. 2 (Forced to land)
7.		06.03.18	Lt. Böning	Rodern	Spad XIII
8.		06.03.18	Lt. Böning	Rodern	Spad XIII
9.		15.03.18	Lt. Böning	Lautenbach	Sopwith 1½ Strutter
10.		15.03.18	Lt. Böning	Lautenbach	Sopwith 1½ Strutter
11.		15.03.18	Uffz. Hopf	Alt-Thann	Sopwith 1½ Strutter

4 The British – A New Opponent

2nd German Army, March 19, 1918 – July 9, 1918

4.1 Liéramont Airfield
March 20, 1918 - April 15, 1918

Above: Lt. Hans Hertrich (left) and his ground crew in front of his "baby-blue" Albatros D Va marked with the white-blue-white spiral ribbon and the personal marking "H". (R. Absmeier)

The Staffel's new operation area was now in the German 2nd Army's sector in northern France. The area of operations extended from Moeuvres near Cambrai in the north, where it bordered the 17th German Army, and ended in the area of Bellicourt in the south, bordering on the 18th German Army area of operations. The reason for the transfer was "Operation Michael", a long-planned large-scale offensive, intended on bringing an end to the war on the Western Front. Following the armistice agreement with Revolutionary Russia, the German Supreme Army Command had substantial resources at its disposal which were transferred from the Eastern Front to the western theatre of operations. Using these additional resources, plans were put in place to breach Allied lines prior to the arrival of larger contingents of American Expeditionary Forces in order to bring about a decision of war in favor of the Central Powers. The Offensive was to be carried out by the 17th, 2nd, and 18th German Armies along a frontline spanning 95 km between Vimy to the north, and La Fére to the south.

The German attack began on the morning of March 21, 1918 at 05:00, while Jagdstaffel 76 was still busy completing preparations in order to resume operations from its new airfield.

Although Flanders and the aerial engagements carried out there against the British Royal Flying Corps stationed there was *"every German fighter pilot's dream"*, as Walter Böning put it, neither he nor the majority of his pilots had any experience in aerial combat against British opponents. Walter Böning's previous unit, Jagdstaffel 19, had been stationed in a French sector until the end of 1917 with only a brief exception. As previously mentioned, the French approach to aerial combat was to take rather a defensive stance, which differed greatly from the British philosophy.

Missions conducted by the Royal Flying Corps were generally characterized by an offensive strategy. Orders of the British "Chief of Staff", Hugh Trenchard for his crews can be summarized as follows: *"Always and under any circumstances, cross the lines, enter German airspace and combat German flying units above their own territory, whenever possible."*

While this strategy put notable pressure on German squadrons, it also resulted in heavy British losses. Neither the French nor the German Air Services would have been able to make up for the losses that the British flyers had to suffer in exchange for carrying out the offensive strategies ordered by their superiors, who hailed from the previous generation. This was only possible by fully exploiting all the resources provided by the British Empire.

On March 19, 1918, the Staffel arrived at the 2nd German Army's front sector; they initially took up position at Neuvilly airfield. However, soon after they arrived there, they received orders to move to Lièramont airfield instead, where Jagdgruppe 5 was formed, consisting of Jagdstaffeln 5 and 46. The commanding officer of the Jagdgruppe was Oblt. Richard Flashar, leader of Jagdstaffel 5.

The airfield was located approximately 26 km north-west of the city of St. Quentin, and about 10 km north-east of Peronne. The Royal Prussian Jagdstaffel 46 had already operated from the airfield for some time; they gladly assisted the newly arrived Staffel move into their quarters on the field. The conditions the Staffel encountered at the new reserve field were no comparison to the well cared-for airfield at Habsheim. As a result of preceding battles, the area resembled a desert. On the airfield, shell craters had only scantily been filled in. When it rained, the airfield turned into a muddy swamp which made taking off and landing impossible.

Right: Lt. Richard Emmerich´s career as a fighter pilot was brief. He arrived on the 25 March 1918, but he was killed in action a little more than two weeks later on 11 April 1918. (R. Kastner)

The aircraft were stored in tents surrounded by trenches in order for the rain to drain, and to keep the ground inside the tents from turning into morass. Accommodation for the pilots and crews was provided by captured British "Nissen Huts" made of corrugated sheet metal which were partially damaged; holes in the walls and roofs were covered with tarp to keep out the rain. The Staffel's main area of operation was the front to the north and south of the city of Albert.

As Walter Böning arrived at Lièramont with his Staffel, his friend Hans Böhning and his Royal Bavarian Jagdstaffel 79 set up at Villeselve airfield in the area of the 18th German Army, located approximately 30 km south-east of Lièramont. As soon as he learned of this news, Walter Böning got in his Albatros D Va and paid a visit to his friend and former Staffel comrade. It was only a short hop of 15 minutes between the two airfields, allowing the friends to pay regular visits to each other during the subsequent period.

It remains to be added that Leutnant der Reserve Richard Emmerich joined the Staffel as a pilot on March 25, 1918. He was from Mutterstadt near Ludwigshafen in the Palatinate region, where he was born on February 7, 1892. Before the war he had been studying to be an Engineer. He joined the military as a volunteer after the beginning of the war. He had been assigned to the 12th Bavarian Feldartillerie-Regiment and was wounded by a shot in his thigh fired by infantry on July 9, 1915. Following initial treatment at field hospital 9 in Comines, he was sent on leave. Upon making a full recovery, he returned to his regiment the same year. On December 16, 1916 he was promoted to Leutnant der Reserve. At the beginning of 1917 he applied to join the Fliegertruppe. This was approved quickly, and he was posted to attend pilot training at Flieger-Ersatz-Abteilung 1 in Schleißheim on March 28, 1917. By mid-April of 1917, he attended pilot training at Bavarian Militär- Fliegerschule 2 in Lachen-Speyerdorf, not too far away from his hometown of Mutterstadt. Following the completion

Left: Oblt. Ernst Angermund was the deputy Staffelführer during Lt. Walter Böning's absence from 29 April to 23 May 1918.

of his training, he was transferred to Armee-Flugpark B on November 1, 1917. Three days later he received orders to transfer to Bavarian Flieger-Abteilung (A) 289. On November 22, 1917, he was transferred to FA (A) 289b. The unit was based at Habsheim airfield and was commanded by Hauptmann Otto Baer. After completing 18 flights against the enemy, he was awarded his pilots badge on January 20, 1918. Motivated by the presence of the Jagdstaffeln based at the same airfield, he volunteered to become a fighter pilot after only two months. On March 6, 1918 he was commanded to Jagdstaffelschule II at Nivelles near Brussels. Following completion of the single-seater course there, he arrived at Jagdstaffel 76 as a newly graduated Jagdflieger.[1]

On March 29, 1918, Unteroffizier Gottfried Stemmler joined the Staffel. He was born on November 20, 1892 in Lautzkirchen, west of Zweibrücken in Saarland. He had opted for a military career even before the beginning of the war and had joined the 17th Bavarian Infanterie-Regiment. When the war began, Stemmler held the rank of Unteroffizier. After only a few short weeks in the field, his legs and chest were seriously injured by shrapnel, requiring lengthy treatment at a field hospital. Following recovery, he was no longer able to serve in the infantry, and transferred to the Fliegertruppe in Schleißheim in the spring of 1915 upon his request. There he was to join newly formed Bavarian Feldflieger-Abteilung 8 as a member of the ground crew. On March 14, 1915 his unit advanced to the Vosges front to join Armee-Abteilung Gaede. After one year, he applied to become a pilot and began pilot training with Bavarian Militär-Fliegerschule 5 in Gersthofen in June of 1917. After successful completion of pilot training he was assigned to newly formed Bavarian Fliegerabteilung (A) 199 which did not transfer to the front until November 5, 1917. The unit initially set up its quarters at Monhofen. His later request to become a fighter pilot was approved. He was commanded to Jagdstaffelsschule II in Nivelles in March of 1918, followed by the assignment to join Jagdstaffel 76. Thus, he was an experienced pilot but a fighter pilot newcomer.[2]

On April 3, 1918, the Staffelführer achieved his first aerial victory in the new operating area despite less-than-ideal weather conditions. Unlike before, he only casually mentioned his success in a letter to his parents:

"Shot down a Sopwith Camel after a dogfight at an altitude of 50 m and about 300 m behind enemy lines."[3]

Unfortunately, further details regarding the location of the kill are lacking, neither can be found in the Staffel war diary transcript nor in Walter Böning's personnel file. In his book, "The Sky, Their Battlefield", Trevor Henshaw allocated Camel D1797 of the 54th Squadron RAF as Walter Böning's opponent (the Squadron reported aerial engagements around 16:50 near Mézières, and at about 17:10 near Rosières involving "Albatros Scouts"); however, the aircraft did not crash in close proximity to the front, but rather made an emergency landing far in the rear areas, which prevented confirmation by German ground observers. This allocation is therefore unlikely to be applicable. The victory was confirmed to Walter Böning as his 14th.

During the following days, additional personnel changes were made at the Staffel. Hans Hertrich and Vzfw. Franz Hüttner transferred to Armee-Flugpark 2.

Not everyone who joined the Fliegertruppe and underwent fighter pilot training was in fact suitable for the job, as indicated in a report written by Lt. Walter Böning a March 31, 1918 referencing the transfer of Lt. Hans Hertrich:

"Lt.d.Res. Hertrich arrived at the Staffel from the Jagdstaffelschule on February 21, 1918. His flying skills do not meet the standards required of a fighter pilot. In addition, his nerves appear to be rather worn-out. He was well-liked and respected by his comrades. His finances are in good order.
Böning
Lt.d.Res and Staffelführer"[4]

Above: Vzfw. Karl Koller. (Koller)

On April 7, 1918 Oberleutnant Walter Angermund and Vizefeldwebel Karl Koller arrived at the Staffel as replacements for the two pilots who transferred. Oberleutnant Walter Angermund graduated from Jagdstaffelschule I while Vizefeldwebel Karl Koller arrived from Jagdstaffelschule II.

Walter Angermund was born on October 15, 1893 in Obersartowitz, West Prussia. He moved to Bavaria and lived in Vilsbiburg, Lower Bavaria. He had joined Bavarian Infanterie-Regiment 6 in 1912 and moved to the front with them at the beginning of the war. In the summer of 1917, he joined the Fliegertruppe; he was initially commanded to Flieger-Ersatz-Abteilung 1 in Schleißheim on September 8, 1917. One month later he attended pilot ground and flight training at local Militär-Fliegerschule 1. Upon completion of pilot training in Schleißheim he was transferred to AFP B and on to Jagdstaffelschule I on March 13, 1918. He subsequently joined Jagdstaffel 76.[5] He was a newbie, both as a pilot and a fighter pilot.

Karl Koller came from Glonn near Munich, where he was born on February 2, 1898. Before the war he spent time in England and worked for the household of a Lord; therefore he spoke fluent English. When the war began, he deployed with the Eisenbahn-Bataillon (Train Bataillon) as a war volunteer. On January 16, 1916, Koller was transferred to Schleißheim where he was assigned to work as a mechanic with Bavarian Flieger-Ersatz-Abteilung 1. He was then commanded to the engine school in Oberursel, which was part of the Oberursel engine factory. From April 4, 1916 until November 16, 1916 he was an engine mechanic with Kampfstaffel 34b. As he desired to be a pilot, he applied for pilot training. Beginning in May 1917, he attended Bavarian Militär-Fliegerschule 4 at Lager Lechfeld. Upon completion of pilot training, he was commanded to Fliegerabteilung (A) 273 on December 4, 1917. Shortly after, he was promoted to Unteroffizier and received his pilots badge on January 14, 1918. The day after his promotion to Vizefeldwebel on February 27, 1918 he was transferred to Bavarian Fliegerabteilung 47 at the Eastern Front. He had volunteered for fighter pilot training in the meantime, and arrived at Jagdstaffelschule II on March 17, 1918 to begin his training there.[6]

Initially, flights were hampered by poor weather conditions, including heavy rain which turned the landing areas into big swamps, along with fog and low clouds. The Staffel's tents were set up in big swampy areas which made keeping equipment dry a challenge for technicians. On April 8, 1918, Walter Böning found the time to write a long letter to his family while sitting in his corrugated sheet metal barrack:

"At the moment the rain is so heavy that we cannot take off, even in our area. I will take advantage of this downtime to write you a letter. Our living conditions here are similar to my time with the infantry in 1915. Five officers share a room; barracks are made from corrugated sheet metal. Luckily, they could not get destroyed too easily (by the retreating British forces, author's note). At last, we are able to make our place a little cozier, albeit slowly. In the long run, a simple mattress made of chicken wire does not meet our spoiled standards. We have secured a few blankets, etc., even a few chairs and heaters. We arrived a few days late to the area, so most of the furniture had already been confiscated by convoys or by one of the 100 or more flying units. Well, my light carriage is operational, providing electrical light in this desert at any rate. The English were keen to get out of here as quickly as possible, and left most of the wells intact, hardly destroyed any of the railway tracks, ammunition, cannons, and left thousands of MG behind. I believe the value of war materials left on the ground to exceed 1 billion, and hardly anyone took any notice of it initially. But the collection crews have been hard at work now; despite their slow progress, their stock is accumulating rapidly.

We'll see in time how things will progress; right now, no one can take the time to catch a breath on either side. Us pilots are doing extremely well, especially once airborne, and even though we have to disengage on occasion, we do feel absolutely superior. I certainly would not like to serve as unit commander on the other side right now, who would only be able to take along half his equipment during the retreat and who would be obliged to burn most of the rest..."[7]

By the evening of April 11, 1918 favorable flying weather prevailed. Around 17:40, six Jagdstaffel 76 aircraft took off in response to a request received from the front, led by their Staffelführer. Their route of flight took them slightly north towards Nurlu, then westbound via Combles towards Albert; observers based there had reported Allied artillery aircraft in the area. When they arrived in the vicinity of Albert, they spotted British R.E. 8 which attacked them from above the clouds.

Lt. Richard Emmerich attacked an R.E. 8 which then attempted to escape in a nosedive. He pursued the aircraft whose observer continuously fired at the German fighter. Suddenly, the Albatros reared up; fragments were seen to detach from the aircraft, and breaking up in the air, it crashed at around 18:20, 4 km west of Albert, where it went up in flames upon impact. After six months at the front, Jagdstaffel 76 had lost its first pilot in aerial combat.

The report of loss for Lt. Emmerich reads as follows:

"At 17:40, Lt. Emmerich departed on a fighter patrol onboard Alb. D Va 5726/17 along with the Staffel leader and four additional aircraft. During the course of the aerial battle, Lt. Emmerich pursued an enemy R.E. to an altitude of 200 m above ground level. Suddenly, Lt. Böning and Uffz. Diem watched as the aircraft's elevator detached, soon after the entire aircraft disintegrated in the air. It is not possible to determine whether the aircraft received a direct artillery hit or if its rigging wired were severed. The aircraft hit the ground at 18:20, 4 km west of Albert behind enemy lines, and burned."[8]

The attacked R.E. 8 aircraft was flown by 2/Lt. R.G. Hart and 2/Lt. L.F. Handford of 15. Sqn. R.A.F. R. Vann & R.C. Bowyer, the other RAF aircrew involved in the aerial combat, reported the

Above: Crash-landing of the black Albatros D III (O.A.W.), formerly the personal plane of Erwin Wenig. The identity of the pilot who took over this machine after Wenig´s departure from Jagdstaffel 76 has not been recorded. By the time of the crash, the first changes from the original Iron Cross markings to the new, straight-sided Balkenkreuz markings had already been carried out, but only on the fuselage and rudder. The Iron Crosses still remained in place on the upper wing. Note the flare pistol opening added to the fuselage side. (T. Weber)

following:

"2/Lts. R.G. Hart and L.F. Handford were flying over low clouds at 3.500 feet (700 meters) over Bouzincourt shortly after 18:00 when they were attacked by four Albatros scouts which had come up through the clouds and fired under the R.E. 8's tail. Hart immediately turned with a view to engaging the E.A. (enemy aircraft) but at that moment his elevator controls were shot away and, a second later, his aileron controls were put out of action. In addition, Hart received a bullet splinter in the ankle. In the meantime, Handford swung his paired Lewis guns into action and opened fire at the nearest Albatros getting about half a drum into it. Both starboard planes of the German fighter broke away and the remains plunged to earth. Bursting into flames the main body hit the ground in British lines near Millencourt.

A second Albatros closed in to avenge his comrade's death and Handford was wounded by two bullets, but before he collapsed, he opened fire on this fresh assailant and the Albatros was seen to fallout of the fight. Subsequent confirmation came from divisions in the line that both enemy aircraft fell in flames. Hart, despite his R.E. 8 being almost uncontrollable brought the aircraft down safely."[9]

The second Albatros D Va which came to Lt. Emmerich's aid was Uffz. Josef Diem's Albatros D Va painted with the insignia of the "Münchner Kindl". He opened fire on the R.E.8. Josef Diem gave chase to the R.E.8, but was obliged to disengage due to heavy British anti-aircraft artillery. Though he had been subjected to heavy fire, he managed to regain altitude and reached German lines unscathed.

Since observers on the German side of the lines had witnessed the fall of the R.E.8, Uffz. Diem was credited with the R.E.8's emergency landing as "forced to land.

Walter Böning was in the process of gathering his aircraft over the lines when a number of S.E. 5a, outnumbering his own formation by a three to one ratio, attacked them by surprise. Uffz. Stemmler's aircraft was singled out by several S.E. 5a from above, and downright shredded by machine gun fire. The aircraft went up in flames and crashed approximately 4 km north of Albert in German lines. Walter Böning, who was the first to spot the attacking S.E. 5a, rushed to his comrades' aid but was unable to prevent his crash. He immediately proceeded to attack the S.E. 5a. The shots he fired hit the British aircraft which crashed, engulfed in flames. In light of being faced with two own aircraft lost, and the numerical superiority of their attackers, the German fighter aircraft had no alternative but to make off as quickly as possible. The S.E.5 disengaged and disappeared back across their own lines.

The loss report for Uffz. Stemmler reads as follows:

"Uffz. Stemmler departed on a fighter patrol on Alb. D Va 7249/17, alongside Lt. Emmerich. During the sortie, 12-14 S.E.'s attacked our aircraft. Uffz. Stemmler was attacked by surprise by one of the S.E. aircraft; he was shot down with just a few rounds and crashed burning at 18:25, approximately 4 km north-east of Albert, in our lines. Lt. Böning immediately attacked the victorious S.E.5 and shot it down in flames.[10]

The S.E. 5a belonged to the 56. Sqn. RAF. The daily RAF „Intelligence Summary" issued the following report about the aerial combat:

"During a general engagement between a formation of No. 56 Squadron and 11 E.A. scouts, Capt. L. W. Jarvis fired about 15 rounds into one of the E.A. A large volume of smoke immediately burst from the E.A. which went down and crashed. This was seen by several pilots of the patrol.

Lt. T. Durrant got on the tail of an E.A. scout and opened fire at rather less than 100 yards range; the E.A. went down in a steep dive and burst of flame appeared and the E.A. was lost to view in a cloud.

K. W. Junor, No 56 Sqn., fired a burst from each gun into an enemy aircraft (E.A.) scout, which was observed to emit puffs of white smoke and water and go down in a steep dive. This E.A. is confirmed to have crashed west of Aveluy village."[11]

There actually were only five German Albatros, rather than the eleven described, which were attacked by the S.E. 5a formation. Uffz. Stemmler was likely shot down by Lt. Trevor Durrant flying S.E. 5a B 181; Durrant was the only Englishman who reported the "burning" crash of his opponent. Apart from Lt. Emmerich, who had been shot down in aerial combat by an R.E. 8 just minutes before, Stemmler was the only German loss. Other German Staffeln did not participate in the battle.

The aircraft which had been shot down in flames by Lt. Böning was not the conqueror of Uffz. Stemmler, but rather the S.E. 5a C5432 flown by 2/ Lt. H J Walkerdine. The British pilot was able to make an emergency landing and jump out of his burning aircraft, in spite of his injuries.

Both of the fallen German pilots had only survived as fighter pilots on the front for approximately two and a half weeks. Their loss was an indication of the great danger these kinds of missions presented for newcomers on the British front.

The following evening, Walter Böning sat in his corrugated-iron hut once again, writing letters on his makeshift knocked together desk. The loss of two of his comrades weighed heavily on him. While he kept his composure in front of his pilots and continued to emit optimism, but silently he kept wondering if he would have been able to prevent their deaths somehow. Finally, he processed his experiences in a letter to his parents:

"Yesterday, trouble was brewing, it was a bad flight. I attacked an R.E. 8 but overshot him. Lt. Emmerich attacked right after me, crosses the line of flight of a heavy projectile and burst into pieces (Author´s note: Böhning´s wording implies he believed Emmerich was hit by an anti-aircraft shell at the time of writing this letter). Shortly after, as there were four of us left, we were attacked by at least 12 S.E. 5a. One of my Uffz. was set on fire. I immediately attacked the S.E. which fired the shots, and he crashed in flames. Then, I got out of there as fast as I could. We lost two, and so did the English. The ratio is off, however. Normally, our losses are 1:3 (at the expense of the British, author's note). The term "air battle" is an expression that people at home seem to have a false concept of. I myself did not know what it was until 3 weeks ago."[12]

In conversation, Dr. Böning shared his thoughts on aerial combat with the Royal Air Force:

"I had never experienced anything like it, swarms of aircraft everywhere from 500 to 4000 meters, the constant rattling of machine guns in the air, aircraft appearing out of nowhere and disappearing just as fast as they came; and there I was, in the middle of it all, trying to keep my people together."[13]

The next day, it was April 12, 1918, at 10:15, British bombers appeared unexpectedly, dropping bombs on the airfield occupied by Jagdstaffeln 46 and 76. The bombs hit tents and squadron quarters. The Jasta 76 ground crew was hit particularly hard. Pionier Friedrich Schlötzer was gravely injured by the detonation of an aerial bomb; he died 20 - 30 minutes after being admitted to the Cappy field hospital. Pioniere Wolfgang Hollweck, Pionier Georg Bär, and Ludwig Lorenz were also injured during the attack; they were transported to Cappy field hospital for further treatment.

On April 14, 1918, Unteroffizier Ludwig Artmann arrived at the Staffel. He was born on July 2, 1897 in Straubing, Lower Bavaria. He worked as a machine fitter before the war and was drafted after the mobilization. He was a member of the 3rd Pionier-Bataillon. In September 1917, he began pilot training with Bavarian Militär-Fliegerschule 2, followed by his transfer to AFP 1at Lachen-Speyerdorf on March 3, 1918. On March 11, he was assigned to Bavarian Fliegerabteilung (A) 286. Following a brief intermezzo there, he was commanded to

Jagdstaffelschule I in Valenciennes. He arrived at the Staffel on April 14, 1918.[14]

On the evening of April 14, 1918 after ground personnel had rectified the damage caused by the bombs to the best of their abilities, Walter Böning sat in his barrack, contemplating. He put his thoughts down on paper:

"I believe today is Sunday; at least I saw a few hundred men line up in front of a real church early this morning, and I heard organ music. The sounds of the music must have appeared to them as though it came from a different world.

Us pilots are the ones who are worst off now, as far as living quarters are concerned; we cannot be any closer to the front. There are no airfields in the cratered grounds, so we end up in the desert. Soon, we will take a risk and advance. In the areas that appear flat and extend about 300 m - 400 m, there are only a little more than 100 grenade craters, and we will be able to fill those. These kinds of areas are rather hard too, however. The two-seater Fliegerabteilungen are fortunate, since they do not have to advance as far forward as we do, or else there would be quite a bit of an argument as to who would get these little spots... My society has shrunk substantially; apart from myself, I am only allowing three other pilots to fly along. Some of the others are unfit for the task, two are sick, the rest have either died in battle, or have transferred elsewhere. With only four aircraft, however, we feel quite worthless. On the front, we are constantly attacked by superior forces, while we hardly get an opportunity to fire a couple of shots ourselves.

I am managing to keep the 'Schweinigel' (dirty pigs, lit: Pig-Hedgehog, German word for a nasty enemy) *at bay, but I worry greatly about my young pilots. I have sent an urgent memo to my superiors requesting immediate replacement of pilots, to include at least one very seasoned fighter pilot who is experienced with flying on the front. I know that Bavarian pilots have been difficult to get, so I added the following: "Should it not be possible to immediately replace the pilots I have requested, I will cease all front flights and request to process my removal from my position as Staffelführer, as I will not be able to assume responsibility for any further losses anymore."*

One has to stick up for oneself, especially within the Bavarian organization. If anyone thinks my tone was too Prussian, or too clear or reckless, I do not care, as long as I will get my pilots. If not, I will resign; I know exactly where I will be welcomed back with open arms. Earlier today, I went to see Rittm. von Richthofen, and I believe that in that kind of environment, I will be able to inflict quite a bit damage to the British in the area. I will see how things go, but I will not take no for an answer.

Otherwise, I am doing great, and I even do believe that all the excitement keeps me on my toes."[15]

The last part of the letter referred to his visit to Jagdstaffel 11 earlier that morning. As it had been raining constantly and no enemy flying activity had been reported, Walter Böning decided to fly from Lièramont to Cappy that morning (approximately 30 km) where Jagdstaffel 11 had arrived two days earlier. He wanted to take a closer look at the Staffel's triplanes, which he had witnessed in aerial combat recently. When he arrived, the triplanes were lined up on the airfield, ready for take-off. However, the majority of the equipment as well as the ground crew, had already relocated to a different airfield in the area of the 6th Army (which was revoked shortly after).

He had introduced himself to Rittmeister Manfred von Richthofen who showed him the Staffel's Fokker Dr. I. He then invited Walter for lunch. During their meal, Walter Böning told his host about the angry letter he had written to the Bavarian Inspectorate of Aviation. Richthofen liked the letter; he remarked that some people at home still fail to fully understand what was going on at the front. He said to Walter Böning:

"With 15 kills I would gladly request your transfer to my outfit."

Manfred von Richthofen then told him that in his opinion the officers in charge at the Inspektion der Flieger had held on to the Albatros D V/D Va aircraft for too long, thereby criminally neglecting the development of new fighter aircraft. He mentioned that Fokker had developed a new aircraft which would make all others pale in comparison, and that he would do everything in his power in order to get these aircraft for his squadron.

In the conversation Walter Böning explained:

"I was deeply impressed by Richthofen's personality, and I am certain that he would have requested me. But as I was part of the Bavarian Army, I am not sure whether it would have actually happened. The discussion with Richthofen was more of an act of defiance on my part; of course, I did not want to let down my pilots, and I was hoping that the bureaucrats in Schleißheim would show some kind of understanding for the situation.

I had a very good pilot, Diem; one good pilot, Hopf; and two pilots, Walther and Bärenfänger; they were experienced but rather war-weary. They had only stayed at the front from a sense of duty because I asked them to stay. Others were quite simply unfit for the job of a fighter pilot. The rest

Above: Tent hangars at Suzanne airfield near Bray. Hand-written notes marked on the print indicate the directions to Bray and Peronne from the airfield. (R. Absmeier)

of them were willing to do the job but they were completely inexperienced. I could not bring myself to send these people into a battle that they were not capable of standing up to. I am not getting my own people killed. Rather than that, they should remove me from my position."[16]

In conversations between Bruno Schmäling and the former German fighter pilots, they frequently addressed the persistent rumor that there was too close of an intertwining between the procurement department of the Inspektion der Flieger in Berlin and the prosperous Albatros company. General consensus was that this had led to the preferential award of large batches of fighter production orders to this company and its subsidiary, the Ostdeutsche Albatros Werke. Fritz Jacobsen, a former pilot of Jagdstaffel 31 and 73, even mentioned that rumors circulated amongst pilots indicating there were financial donations from the Albatros Company made to people in certain positions within the Inspektion der Flieger, in order to decide in favor of the Albatros company. One name that was repeatedly mentioned in that context was that of Hptm. Kurt Schwarzenberger, who was responsible for the technical development of fighter aircraft with Inspektion der Flieger; he was said to have maintained a very close relationship with the Albatros Werke. Of course, after the passage of more than a century, it is no longer possible to actually prove to what extent these rumors were true. However, it is interesting pointing out that the same rumor had been shared with Scottish aviation historian Alex Imrie when he interviewed a number of former German fighter pilots during the 1950s and 60s.

4.2 Suzanne Airfield April 15, 1918 – July 9, 1918

Due to the advances made during the German Offensive the Staffel relocated to Suzanne reserve airfield in the pouring rain on April 15, 1918. The new airfield was located approximately 22 km west of their former airfield in Liéramont, north of the Somme River, and not too far from the village of Bray sur Somme, which had been battered severely during the recent fighting. A certain amount of fighting had occurred here, too, but the airfield and the area surrounding it had suffered less than it had been the case at Liéramont.

Cappy airfield occupied by Jagdstaffel 11 was located just 2 km south-west, on the other shore of the Somme River. This was reason enough for Walter Böning and his pilots to visit Jagdgeschwader I on occasion, where they admired the Jagdstaffel 11 triplanes.

As Jagdstaffel 79 had already set up their quarters with the German 18th Army at Ercheu airfield on April 7, 1918 he was able to keep in touch with his friend, Hans Böhning. Ercheu is located only 40 km south-east of Suzanne, allowing the two friends to pay frequent visits to each other.

As before, the Staffel's main operating area included the region north of Albert, but their missions now included regions south of Amiens, abeam Boves.

On April 20, 1918 orders arrived from Bavarian Inspektion des Flugwesens, issuing transfer orders for Vzfw. Siegfried Walther to proceed to Flieger-Ersatz-Abteilung 1 at Schleißheim by April 29, 1918.

With Vzfw. Walther's departure, the Staffel lost another experienced and brave Jagdflieger. He was transferred to Armee-Flugpark F in Palestine on June 21, 1918 from where he proceeded to Jagdstaffel F 1 or F 2. He fell in August 1918 serving with his new unit, but available information regarding the exact time and circumstances of his death is contradictory.[1]

When transfer orders arrived for Uffz. Karl Hopf on the same day, Walter Böning was absolutely furious. He found himself now left with just a single capable pilot, Uffz. Josef Diem, as well as worn out Lt. Karolus Bärenfänger. The rest of his pilots consisted of inexperienced newcomers. It appeared that some high-ranking officers in the Bavarian Inspectorate of Military Aviation wanted to prove to the rebellious Leutnant Böning who was really in charge of things. His resentment of the responsible officers serving with the Bavarian Inspektion des Flugwesens had reached a new high:

"Instead of receiving an experienced pilot, two of them are taken from me! How am I supposed to face the British with that kind of team? Usually, there were no more than four or five of us in the air at a time, as we could only keep a watchful eye on two newbies at a time. We were under attack soon after getting airborne; I spent more time protecting my boys than attacking the enemy. The qualified pilots required for our own Bavarian Jagdstaffel were most definitely missing, and it was irresponsible to form our own Bavarian Jagdstaffeln under these circumstances!"[2]

Uffz. Hopf was transferred to Jagdstaffel 80 via Armee-Flugpark 2, Bavarian Flieger-Ersatz-Abteilung 1, and Armee-Flugpark 19. Oblt. Erwin Wenig, Staffelführer of Jagdstaffel 80, had urgently requested an experienced fighter pilot. Apparently, he had found himself in a position identical to Walter Böning. Erwin Wenig, who did not request Hopf specifically, later contacted Walter Böning by telephone, apologizing to him for taking away one of his best men. This kind gesture was well received by

Above: Lt. Edmund Neu arrived at Jagdstaffel 76 on 23 April 1918 and was wounded in action on 16 July 1918. He returned to the Staffel on the 12. August 1918. (R. Kastner)

Above: Lt. Hermann Obermeier served with the Staffel until the end of the war. (R. Kastner)

Böning, as he and Erwin Wenig were good friends. He continued to love Bavaria and its people; however, the ties he had with the Bavarian Army had now been permanently severed. On a side note, Karl Hopf arrived at his new Staffel on July 27, 1918 (likely a result of Erwin Wenig's urgent request).

One day later, a shocking message from Jagdstaffel 11 reached the Staffel: *"The Rittmeister is missing!"*.

"We were all literally paralyzed; I had witnessed Richthofen several times, his dynamism, his personality, his will... now, he was no longer. He was one of the most impressive people that I had ever met in my life."[3]

Between April 20 and 23, 1918 three new pilots arrived at the Staffel in order to replace the experienced Vzfw. Walther and Uffz. Hopf, who had recently departed. The new pilots were Leutnant der Reserve Moritz Weig, Leutnant der Reserve Edmund Neu, and Leutnant der Reserve Hermann Obermeier. They were all "Häschen" (literally: bunnies - newcomers) that arrived from Jagdstaffelschule I. Their presence did not immediately increase the Staffel's fighting power but rather weakened it. Now, the few remaining experienced pilots had to watch over even more new Staffel pilots.

Moritz Weig was born in Wallhofen on September 22, 1891; at the beginning of the war, he was a merchant living in Gräfelfing near München with his parents. He reported as a war volunteer and served with the Stationary Motor Vehicles Depot (Immobiles Kraftwagendepot) 6 in Munich. On October 10, he deployed to the front as a driver with Bavarian Feldflieger-Abteilung 4. Following his transfer to the Bavarian 12th Reserve-Infantry-

Above: Test of one of the newly delivered parachutes at the beginning of May 1918. Systematic equipment of the German flying crews with parachutes began in spring of 1918.

Regiment and a slight injury, he was promoted to Leutnant der Reserve on September 20, 1916. In the spring of 1917, he volunteered for service with the Fliegertruppe and was commanded to Bavarian Flieger-Ersatz-Abteilung in Schleißheim on August 15, 1917 where he was to attend pilot training with the local Fliegerschule 1 eight days later. Upon completion of pilot training, he was transferred to Armee-Flugpark 17 on April 12, 1918. Two days later, he was commanded to Jagdstaffelschule I. From there, his way led him to join Jagdstaffel 76 on April 22, 1918.[4]

Edmund Neu was from Oggersheim in the Palatinate region where he was born on March 9, 1896. He started the war as a volunteer with the Bavarian 23rd Infanterie-Regiment, where he was promoted to the rank of an active Leutnant on March 10, 1916, and he joined the Fliegertruppe in 1917. On April 2, 1917 he was commanded to Bavarian Flieger-Ersatz-Abteilug 1 in Schleißheim. In mid-May 1917 he began pilot training with Militär-Fliegerschule 2. Upon completion of pilot training and after his transfer to Armee-Flugpark A, he was commanded to Jagdstaffelschule I. He arrived at Jagdstaffel 76 on April 23, 1918.[5]

Hermann Obermeier was born on September 16, 1893 in Elberfeld, which today is part of the city of Wuppertal. Before the war he was studying to become a lawyer. He signed up as a war volunteer and was assigned to Bavarian 2nd Fußartillerie-Regiment. He was promoted to Leutnant der Reserve on March 11, 1916. Following his request to join the Fliegertruppe, he attended initial training with Flieger-Beobachterschule Königsberg in Sepember 1916, followed by further observer training with FEA in Schleißheim. On February 1, 1917 he was transferred to AFP 1 and assigned to Fliegerabteilung (A) 286. On June 22, 1917 he was transferred to Schleißheim as an instructor at Bavarian Flieger-Beobachterschule 1. On October 30, 1917 he transferred to Bavarian Militär-Fliegerschule 2 to attend pilot training. On April 15, 1918 Lt. Obermeier was transferred to Armee-Flugpark 2 and later commanded to Jagdstaffelschule I, where he attended fighter pilot training for several days. On April 23, 1918 he transferred to Jagdstaffel 76.[6] Although he was not an experienced fighter pilot, he had gained some front flying experience during his time as a pilot with Fliegerabteilung (A) 286.

Upon arrival of the new pilots, Walter Böning asked Karolus Bärenfänger, his most experienced pilot, to take charge of familiarizing the "Häschen" once again with their new environment and keeping a close eye on them on missions. Lt. Bärenfänger had seen uninterrupted service as a fighter pilot on the Western Front as a fighter pilot since March

Above: Lt. Moritz Weig's Albatros D Va came down in a "Damenlandung" behind French lines on May 2, 1918 following aerial combat with S/Lt. Marcel Coadou of Escadrille 88. (Musée Air + Espace, Le Bourget)

Below: Lt. Moritz Weig's Albatros D Va came down in a "Damenlandung" behind French lines on May 2, 1918 following aerial combat with S/Lt. Marcel Coadou of Escadrille 88. (R. Absmeier)

Above: Adj. Marcel Coadou of Escadrille Spa. 88as a student pilot. He was, Lt. Moritz Weig's opponent in aerial combat on 2 May 1918. (D. Méchin)

1917. He had reached the limit of his physical and psychological capabilities, which meant that he frequently had to take recovery breaks between flights.

Walter Böning remembers, *"...he of all pilots would have been most entitled to a replacement, but he chose to stay out of a sense of obligation and as a result of my request to remain with the Staffel."*[7]

Initially, the morning of April 23, 1918 was misty with poor visibility. However, as the day progressed, the weather cleared. Enemy flying activities did not markedly increase until the afternoon, reaching its peak towards the evening. Just before 19:00, a violent air battle unfolded in the vicinity of Sailly-Laurette between British fighter aircraft and German single-seaters from Jagdgeschwader I.[8] Some Jagdstaffel 76 aircraft were also engaged in this encounter. Walter Böning later reported:

"In a joint effort with others, I shot down an English S.E. which caught fire and disintegrated. The other guy received credit for the kill, which was probably the right decision to make. During the course of this single-seater encounter, there were about 20 Englishmen against 20 Germans, I received two hits. One of the shots shattered the main spar, the other, a phosphorous bullet, pierced the horizontal tube of the elevator and ricocheted through the fuselage into the pilot seat, tore open my coat, and the phosphorus charred my sheepskin. Of course, I felt the impact and got a bit upset, but as I felt no blood and only found the bullet along with a few lead fragments (which were so kindly encased by the Brits), I quickly regained my calm..."[9]

The aircraft shot down was S.E. 5a C1086. Its pilot, Capt. Kenneth William Junor from 56. Sqn. RAF, did not survive the crash. He had crossed blades with Jagdstaffel 76 previously, on April 11, when Uffz. Stemmler was shot down. This time he had pulled the shorter straw, proving just how closely related triumph and tragedy were in times of war. Walter Böning was not the only one putting in a claim for this particular kill, and eventually confirmation for it was granted to Lt. Egon Koepsch of Jagdstaffel 4.

Further details relating to this encounter can be found in Alex Revell´s formidable history of 56. Squadron RAF *"High In The Empty Blue"*:

"On the morning of April 23, rain and low cloud stopped all flying; but the skies brightened at mid-day - although the cloud remained - and ten SEs took off at 4:30pm, led by Galley and Hastings-Trew. Galley, Junor, Irwin, Parry, Macdonald, and Doyle attacked five Albatri over Brecourt at 5:00. Galley's opponent spun away from his fire and Parry's spun for several turns before straightening out and 'lurching about.' The remaining Albatri made off east and the SEs reformed and flew north. Galley and Junor next attacked a two-seater over Achiet-le-Grand, but the enemy pilot dived into the mist and escaped. At 6:00, 'B' Flight attacked five Albatri east of Bray, but seeing ten Fokker Triplanes approaching from the south, with a height advantage of 2,000 feet, the Flight broke off the action and climbed north-with the exception of Junor, who continued to fight the Albatri. Hastings-Trew, Garrett, and Stenning, flying top cover for 'B' Flight, had seen the Triplanes arriving and dived to attack them. Hastings-Trew's Vickers gun jammed and his Aldis sight fogged up during the dive and seeing that Galley and 'B' Flight had evaded the Triplanes' attacks he pulled out of his dive and zoomed away, followed by Garrett and Stenning. A formation of SEs from another squadron then

Above: S/Lt. Marcel Coadou of Escadrille Spa. 88 in one of his Spad VII with the name "*Judex*". *Judex* was a twelve-part popular French silent movie from 1916 about one of the first disguised avengers in film history. (D. Méchin)

arrived, reinforced by a number of Sopwith Camels, and began to fight the Triplanes.

Over Fricourt, Galley later attacked an Albatros D.V. sporting a light blue tail. The enemy pilot dived away east, but Galley followed, firing an entire drum of Lewis and 100 rounds of Vickers. The Albatros suddenly went into a perpendicular dive and disappeared into the low lying mist. Galley was awarded this Albatros as out of control. Galley then reformed the Flight and returned.

On landing it was found that Kenneth Junor had failed to return. He had last been seen fighting the Albatri and Triplanes and it was later learnt that he had been shot down and killed. Junor had been promoted to Captain on April 21 and given command of 'A' Flight, replacing Captain Billinge, who had been posted to Home Establishment. The Squadron could ill afford Junor's loss for he had shown great promise as a potentially able Flight Commander. Junor had been shot down by Ltn. Koepsch of Jasta 4. The RAF casualty returns for April 23, 1918 show no losses in the morning there was very little flying owing to the bad weather conditions and Junor's was the only SE5 lost by the RAF that day. The War Diary of Jagdgeschwader 1 gives Koepsch's victory, an SE5, as having crashed in the vicinity of Sailly-Laurette at 6:55vorm (am). With the exception of this 12 hour time difference, these facts, place and time (allowing for the one hour time differential in German time) fit the circumstances of Junor's loss, and it seems probable that a mistake was made in the War Diary of JG I, giving a. m. instead of p. m."[10]

Without a doubt, the light-blue-tailed Albatros D Va attacked by Galley and credited to him as out-of-control was a plane from Jagdstaffel 76, but the unit recorded no losses resulting from this encounter.

During the following days through the end of the month, poor weather conditions prevailed, bringing fog and low clouds to the area which significantly impacted the missions of the flying units, finally prohibiting all flying activities between April 26 and April 30, 1918.

On April 29, 1918 Walter Böning went on his long approved three-week home leave. Oblt. Ernst Angermund, the highest-ranking officer in the

Above: S/Lt. Marcel Coadou and a comrade pose in front of his SPAD VII *"Judex"*. (D. Méchin)

Staffel, would serve as his deputy in his absence. Angermund was a reliable and well-liked officer who, like Walter Böning, could have picked a quieter Oberleutnant assignment rather than being a fighter pilot. However, he wanted to do his part as a flyer serving Germany in the war. Although a skilled and courageous flyer, putting him in charge of leading the Staffel was a risky decision, as he only had 4 weeks of frontline experience as a fighter pilot at the time. Therefore, Walter Böning reverted to a method commonly used in fighter units; nominally, the Staffel was led by Oblt. Angermund, which met the requirements set forth by higher Staff members who would not accept anyone other than an officer as their point of contact in his absence. Oblt. Angermund was in charge of all day-to-day ground operations. However, when carrying out aerial sorties, it was up to the most experienced and best pilot to lead the Staffel - Uffz. Josef Diem.

Walter Böning had pulled Ernst Angermund aside prior to his departure and made it unambiguously clear that he was to abide by his wishes. Of course, this was quite unusual of a measure considering the large difference in ranks between the two, but Oblt. Angermund was fully aware of the motivations his comrade was driven by and accepted.

Finally, Walter Böning inculcated in both of them not to take any unnecessary risks, and above all, to get "the boys" through the following three weeks unharmed. As he later recalled, he left for his home leave with an odd, strange feeling. However, after spending such a long period at the front he was in urgent need of some rest. While he was physically away from the Staffel, his thoughts frequently were with his men on the front.

At the beginning of May 1918, the first examples of the new parachutes were supplied to at Jagdstaffel 76. They were greeted with much skepticism at first; initially, the parachutes were tested with attached sandbags by dropping them out of an aircraft. After the parachute opened without any issues, one of the pilots, Josef Diem, got into the rear seat of a two-seater, took off and departed the airplane by taking to the parachute, which opened without any problems, allowing the pilot to safely return to the ground. From that moment, the pilots trusted their parachutes; none of them would get into an aircraft without wearing their newly issued "lifesavers". Walter Böning reported that the flyers were prepared to take higher risks with the parachutes on board of their aircraft than before, knowing they had a way of getting out of the aircraft and survive, should the need to do so occur.

On the following day, May 2, 1918, one of the newly arrived "Häschen", Lt. Moritz Weig, did not return from a frontline flight. He had departed at the controls of Lt. Bärenfänger's old, white and blue striped Albatros D Va as part of a Kette which was heading towards the front at Amiens. During the flight, he lost sight of the other aircraft. The Kette returned to the Staffel after a largely uneventful flight, assuming that Moritz Weig had returned due to technical issues. Upon arrival, they discovered that he had not returned home.

Offizier z.b.V. Clemens Vollert immediately began making phone calls in an effort to find out the whereabouts of Moritz Weig. There was hope that he may have landed somewhere on the German side. Initially, his investigations were unsuccessful. Eventually, an air defense officer sent word of a German single seater disappearing overhead enemy territory. The office clerk responsible for making entries in the unit´s war diary noted:

"Lt. Moritz Weig departed on a front flight and did not return. He likely got lost and eventually made an emergency landing on enemy territory."

Moritz Weig had indeed lost contact with his Kette in the vicinity of Amiens and found himself alone above enemy ground. He was most certainly overwhelmed by the situation. Instead of keeping his cool and flying eastwards in order to get back across the frontlines, he ended up flying further to the south south-west into the enemy rear area in a failed attempt to find a familiar reference point.

According to French documents, he was attacked by pilot Adjt. Arthur Marie Marcel Coadou of Escadrille 88 in French territory between Dompierre and Godenvillers, approximately 8 km south of Montdidier, while coming under fire from anti-aircraft guns at the same time. The Albatros D Va went down in French lines, flipping on its back after touching down. The site of the crash is an indication of just how disoriented Moritz Weig had become. He crashed approximately 36 km south of Amiens where he had lost contact with the Staffel, as well as approximately 20 km west of the German front on French territory, near Lassigny.

The following short note on the occurrence can be found on a website about French fighter operations in WW I:

"Le 2 Mai, il attaque, toujours dans la région de Montdidier un biplace, puis un monoplace Albatros qui s›écrase dans les lignes françaises entre Dompierre et Godenvillers. Revenu à Fouquerolles (Oise), terrain de la 88, il saute dans une voiture pour aller sur les lieux du crash. Arrivé sur place, il constate la mort du pilote allemand mais aussi que la D.C.A. des environs s›est attribué la victoire !! Toujours pas d'homologation !"

*"On May 2, he (Marcel Coadou) attacked a two-seater in the Montdidier region when a single seater Albatros crashed in French lines between Dompierre and Godenvillers. Back in Fouquerolles (Oise), the airfield of Escadrille 88 he got into a car in order to travel to the site of the crash. Upon arrival, he noted the death of the German pilot, but he also recorded that the kill had been awarded to anti-aircraft artillery nearby! Still, no recognition! (*Of this aerial victory, author's note*)"*[11]

The airfield from which Escadrille 88 operated, Fouquerollers, was located approximately 30 km from the site where Moritz Weig made an emergency landing. This means that the French pilot first had to return for landing at the airfield before he could get into a car in order to return to the site of the crash. It is assumed that he arrived approximately 45 minutes after the emergency landing. In the meantime, Moritz Weig had already been taken away which may have led Marcel Coadou to assume that the German pilot had died in the crash, which was not the case. Moritz Weig had already been taken away and from the crash site and was released from French war captivity on February 14, 1920.

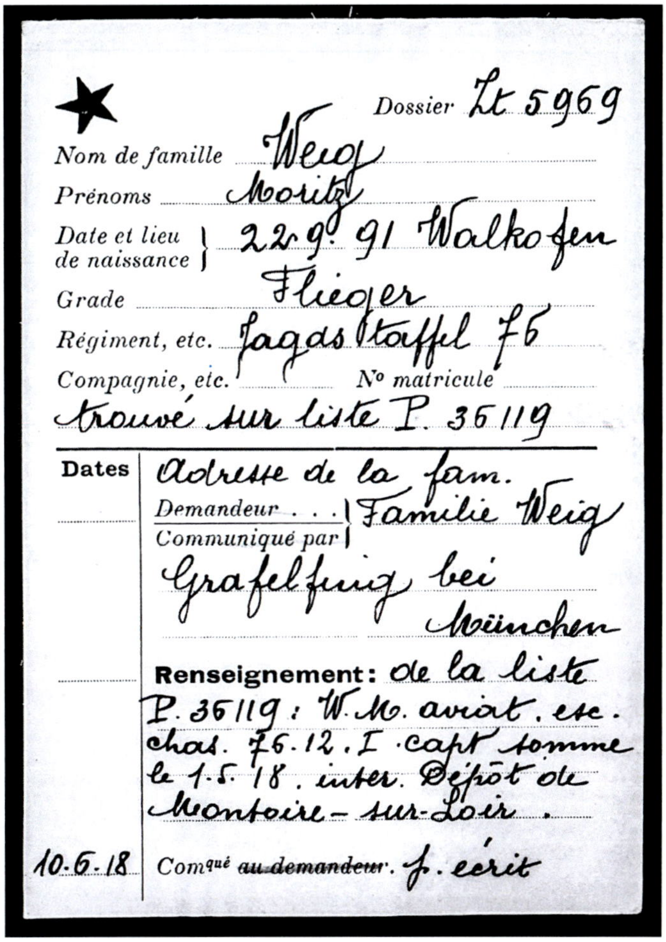
★ Dossier Lt 5969

Nom de famille Weig

Prénoms Moritz

Date et lieu de naissance 22.9. 91 Walkofen

Grade Flieger

Régiment, etc. Jagdstaffel 76

Compagnie, etc. N° matricule

trouvé sur liste P. 35119

Dates | adresse de la fam.
Demandeur . . . / Communiqué par | Familie Weig
Grafelfing bei München

Renseignement: de la liste P. 35119: W. M. aviat. esc. chas. 76.12. I capt Somme le 1.5.18. inter. Dépôt de Montoire-sur-Loir.

10.6.18 Comqué ~~au demandeur~~. f. écrit

Above: Lt. Moritz Weig´s POW registration card. Noted on the card are several of his personal details, including date and place of birth, as well as his unit, Jagdstaffel 76, noted in German language. However, the date of his capture is wrongly recorded as May 1, 1918. (http://pilotesdechasse)

On the same day, Uffz. Josef Diem was finally awarded his pilot badge following two and a half months of duty as a fighter pilot, and multiple requests entered by the Staffelführer.[12]

Two days after Josef Diem received his pilot badge, two more new pilots arrived at the Staffel: Leutnant der Reserve Franz Martin and Vizefeldwebel Georg Markert.

Above: At Suzanne airfield. From left: unknown, Lt. Franz Martin, Oblt. Ernst Angermund, and Lt. Karl Neu. (R. Kastner)

Leutnant der Reserve Franz Martin was from Lower Franconia. His place of birth was Schwärzelbach, a small town in the Südrhön region near Hammelburg in the northwest of Bavaria. He was a teacher and joined the Bavarian 9th Infanterie-Regiment as a one-year volunteer at the beginning of the war. In March 1915 he was promoted to Leutnant der Reserve and Zugführer der 3. Kompanie. In February 1917, Martin was commanded to Bavarian FEA 1. On March 6, 1917 he was transferred to Militär-Fliegerschule 2 in Lachen-Speyerdorf, where he attended theoretical and practical pilot training. Upon successful completion of his training, Lt. Martin was commanded initially to Bavarian Fliegerabteilung (A) 294 on September 8, 1917; two days later, official orders for his transfer to the Abteilung were issued. On December 1, 1917 Martin received a shot to his shoulder during aerial combat. He was taken to a field hospital after landing.

After his recovery, he shot down a Nieuport in aerial combat near Warneton on January 9, 1918 along with his observer Lt. Richard Altvater, receiving credit for his 1st aerial victory. On January 22, he received the honorary cup "Dem Sieger im Luftkampf" ("To the victor in aerial combat"); on February 25, he was awarded the E. K. (Iron Cross) 1st Class. Shortly thereafter, he was wounded again, this time during an engagement with five Sopwith beyond enemy lines between the Ploegsteert forest and Armentières on February 17, 1918, Lt. Martin received a minor grazing shot to his upper lip. The successful air combat on January 9, 1918 may well have motivated him to volounteer for service with the single-seater units upon his recovery. From March 19, 1918 he attended fighter pilot training, initially with Jagdstaffelschule II and later with Jagdstaffelschule I. From there, he was transferred to Jagdstaffel 76. He was courageous and slightly reckless, but among the fighter pilots he was another novice.[13]

Georg Markert was also Franconian. He was born on May 14, 1892 in Wallmersbach, Lower Franconia. He was a professional soldier with the Bavarian 5th Feldartillerie-Regiment, lived in Ansbach and was married with one child. At the outbreak of the war, his Regiment was relocated to the front; on May 11, 1915 they were awarded the Bavarian Silver Tapferkeits-Medaille (Medal of Bravery).

In 1917 he signed up with the Fliegertruppe and was commanded to the Bavarian Flieger-Ersatz-Abteilung in Schleißheim on September 7, 1917. The same month he began pilot training with Militär-Fliegerschule 6 in Bamberg. Upon graduating from pilot training, Markert was transferred to AFP 6 on April 17, 1918. Soon after, he attended a single-seater course at Jagdstaffelschule I. On May 6, 1918 he began his duty with Jagdstaffel 76.[14] He lacked any kind of experience as a frontline pilot.

On May 6, 1918, Leutnant der Reserve Max Schick arrived at the Staffel. He was born in Hamburg on May 1, 1894. He worked as a court assistant in Munich where his parents ran a letterpress. When the war began, he deployed as a volunteer with the 1st Bavarian Infanterie-Regiment. He was wounded twice during the war, and finally, on February 15, 1917 he was promoted to Leutnant der Reserve. He enlisted with the Fliegertruppe and was commanded to Bavarian Flieger-Ersatz-Abteilung 1 on September 21, 1917. He completed pilot training with Militär-Fliegerschule 7 in Germersheim. Following his temporary assignment to Jagdstaffelschule I and single-seater training there, Lt. Schick was transferred to Jagdstaffel 76.[15]

The following Staffel missions did not yield any definitive successes, but there were no losses either, which was a small success considering the fact that the Staffel consisted almost exclusively of newcomers. On May 11, 1918 Lt. Hermann Obermeier made an emergency landing, likely as a result of aerial combat. He severely injured his left ankle and was brought to field hospital 90. On June 3, he was transferred to Bavarian Flieger-Ersatz-Abteilung Schleißheim, but he returned to the Staffel on July 23, 1918.

The only Staffel success during Walter Böning's absence was achieved by Vzfw. Karl Koller. Staffel aircraft had attacked a group of D.H. 4 two-seaters over Allied territory north-west of Mesnil on May 17, 1918. In spite of heavy defensive fire, Vzfw. Koller managed to score decisive hits on the aircraft, setting it on fire. According to German observations, the aircraft crashed in British lines. The kill was Vzfw. Koller's first aerial victory.

Walter Böning returned from home leave on May 20, 1918; he was relieved that his worst fears had not come to pass. Apart from Lt. Weig, who had made an emergency landing on enemy territory after losing orientation, and Lt. Obermeier´s aforementioned injury following an emergency landing, all of his "Häschen" managed to survived his absence unscathed. He thanked Ernst Angermund and Josef Diem for their cautious Staffel leadership. He had not expected any successes under these circumstances. He was impressed with Karl Koller's aerial victory as he was a newbie, and planned to take him under his wing.

Above: Clemens Vollert showing the spoils of his successful fishing day on 20 May 1918.

The following day, word reached the Staffel that Walter Böning was to be awarded the königlicher Hausorden von Hohenzollern mit Schwertern (Royal Hohenzollern House Order with Swords) in recognition of his military achievements. This was reason enough for a small celebratory party.

On the same day, he wrote to his parents:

"I have been here since yesterday afternoon. Our entire Staffel is now resting, which has been long anticipated and necessary. We only fly in order to conduct gunnery exercises and to practice. I am living in a so-called bomb-proof shelter, more than 40 railroad tracks and many cubic meters of sand are above me. Not much has changed here; two of my pilots had one success each. We went fishing

Above: Oblt. Ernst Angermund's Albatros D Va which he crash-landed after being wounded on 23 May 1918. The force of impact tore out the engine with its mounting.

again today, nothing too challenging. Our job involved raising the gillnets which had been set up last night. Among others, we caught two large pikes, many lampreys, and eels. It has been so hot here that we should be dismissed from our duties due to excess heat; I am glad to be sitting in the cellar."[16]

Walter Böning grew increasingly concerned with the hopelessly inferior aircraft his unit was to operate with. The Albatros D Va were not able to keep up with the maneuverability and climbing performance of the current Allied fighter aircraft. Former pilots of other Staffeln agreed that they were no longer able to reach Allied reconnaissance and bomber aircraft in their Albatros or Pfalz aircraft. Additionally, Allied fighter aircraft had the ability to attack from above due to superior climb performance, and to return to higher altitudes following the attacks. There, they were able to regroup and begin another attack while the Albatros and Pfalz aircraft below were flying along as if being presented to the enemy on a silver platter.

Walter Böning had experienced the quality of the Fokker Dr. I of Jagdgeschwader I in combat; he fervently wished to add them to his Staffel aircraft inventory. It was a wish that would obviously not come to pass.

The duty rest periods implemented for the Staffel on May 8 and May 20, 1918 respectively were of short duration. Jagdstaffel 76 returned for duty on May 23, 1918 upon requests issued from the front.

During their first sortie on that sunny day, two Ketten of the Staffel encountered S.E. 5a from 60. Sqn. RAF near Albert. During the ensuing combat, Oblt. Angermund was wounded by a grazing shot to his knee. He entered a spin in an effort to lose altitude, trying to make an emergency landing in a field near Fricourt. In the flare, the aircraft suddenly turned on its side and dropped rapidly, hitting the ground hard. The right wing was torn off, causing the Albatros to slide into a ditch where it turned over, ripping the engine from its anchoring. With shock, Walter Angermund bit his tongue as well. German soldiers rushed to his aid, recovered him from the wreckage, and drove him to the closest field hospital.

His opponent in the air was Capt. J.D. Belgrave of 60. Sqn. R.A.F. who reported:

"... I saw some Albatros Scouts, several miles over the lines. I flew towards them but as we were below their level, I did not attack and allowed them to chase us to the lines. Soon afterwards they dived towards black Anti-Aircraft bursts south of Albert. I then flew to the east and attacked one. I fired and he turned east, then west and I got on to his tail and fired both guns until within about 5 yards of the enemy aircraft. I actually saw bullets hitting the fuselage and pilot and suddenly he fell back and the machine fell over to the right. I dived under the tail missing his undercarriage by a few feet. Lieut.

Above: Vzfw. Karl Koller's Albatros D Va D.7221/17; he was taken into British captivity on 25 May 1918; his aircraft was seized by the British. Koller was brought down by Canadian Alfred Alexander Leitch, who flew with 65. Sqdn RAF. (Koller)

H.A. Saunders followed him down some distance and was then attacked by the remaining 5 enemy aircraft. I dived at these enemy aircraft who drove Lieut. Saunders to Albert but as soon as I opened fire, they retired east. The Albatros crashed near Fricourt."[17]

Walter Böning wrote about the accident in a letter to his parents on May 24, 1918:

"I do not believe that I have already shared with you the reason for my having to be here. Oblt. Angermund was lightly wounded in aerial combat; he had to make an emergency landing. His aircraft flipped over, and he bit his tongue, hard. He was taken to the field hospital. The other officer broke his ankle (Lt. Hermann Obermeier, author's note), so I guess I had to show up. My catarrh is almost over even though I was only able to fly twice so far, it has been raining substantially. No news otherwise; tonight, I will go on a duck hunt; yesterday (when I did not have my rifle) more than 100 ducks flew up from the reed 20 steps in front of me."[18]

The next day at approximately 20:00, Vzfw. Karl Koller departed on Albatros D Va D.7221/17 under overcast skies, accompanied by two Staffel comrades. They were to carry out a patrol flight from which Koller did not return. As it became known later, he had made an emergency landing on the Allied side following an aerial encounter with British fighters and was captured.

His opponents were a Flight from 65. Squadron RAF who wrote the following report upon their return:

Lt. Eric Guy Brookes on Camel C8291:

"While flying over Albert I observed three Albatros Scouts attacking a R.E. 8 with four more Albatros waiting above. I attacked the 3 Albatros and fired a burst into the nearest. Some pieces fell off him and he stalled and went down vertically."

Lt. Joseph Leonard Maries White on Camel C8272:

"A Flight attacked 7 enemy aircraft over Albert at 7:40 p.m. (British time) at about 4000 feet. I attacked one which was on the tail of a Camel, got to very close range and fired a short burst. Enemy aircraft did one turn of a spin and went down in a nosedive. I followed and fired another short burst and observed enemy aircraft to crash S.E. of Albert."

Lt. Alfred Alexander Leitch on Camel D1903:

"While leading an offensive patrol I saw an Albatros Scout over Albert at about 3000 feet at 7:50 p.m. Pursued enemy aircraft which went north-west, opened fire at about 200 yards, enemy aircraft crashed near Senlis."[19]

In the aerial encounter described in these three reports, Jagdstaffel 76 only lost one aircraft rather than three. Vzfw. Karl Koller's opponent was Lt. Leitch; he was the only one reporting that his enemy's aircraft had crashed in British lines. It seems possible that the Jagdstaffel 76 Kette was surprised by 65. Squadron while they were attacking an R.E. 8 which caused their formation to break up; Lt. Leitch only reported an aerial battle involving one single Albatros. However, it is also plausible that the three Albatros attacking the R.E. 8 were aircraft from a different German unit, and that Karl

Above: Vzfw. Karl Koller's dismantled Albatros D Va D.7221/17 is taken away by British soldiers on a truck following his emergency landing. A Sopwith Camel of 65. Squadron RAF marked with the letter "H" can be seen in the background. (G. VanWyngarden)

Koller lost contact with his Kette without enemy interference. He would then have attempted to return home alone, thereby encountering Lt. Leitch. The captured Albatros was salvaged by the RAF and filed under reference number G/3Brigade/7 in the list of captured German aircraft.

Several weeks later, Offizier z.b.V. Clemens Vollert received a letter from Karl Koller, written by him while he was in a British POW camp. The letter was relayed to Jagdstaffel 76 via the Red Cross:

"Sunday, July 7, 1918

Herr Leutnant. I have been in England for some time now. I am doing ok, but I still cannot get over my misfortune. Here, I am forced to waste my time doing nothing while I could easily enjoy some great successes over there. If only I could fly for some 1.5 hours once again, I would give a lot for it. What happened to my beautiful machine is truly a shame, I really smashed it to pieces. I wonder how my Staffel is doing, I hope you have enjoyed a lot of successes? What wonderful times we had by the Somme river. I wonder if my Ehrenbecher (honorary cup) arrived?

I have written a letter to Herrn Leutnant Böning as well. Hopefully my letters will find their way to you. I would very much like to hear a few lines from Herrn Leutnant as I am here in my misfortune. Passing on my best wishes to Herrn Leutnant and the rest of the Staffel men.

I remain Herrn Leutnant's humble servant, K. Koller"[20]

On May 28, an aerial battle was fought between the Albatros fighters of Jasta 76 and S.E. 5 aircraft from 84. Sqn. RAF on a cloudy day. Vzfw. Josef Diem succeeded in forcing an S.E. 5 to land near Hangard in British lines. He was only credited with this victory as an enemy aircraft "forced to land behind enemy lines". Lt. Franz Martin sustained a minor injury during the same encounter; he was able to return to Suzanne airfield and landed safely.

Two days later, against all odds, Lt. Walter Böning enjoyed his first success after returning from his recent leave. His flying skills and experience were enough to make up in part for the Staffel's material inferiority. He had departed on a flight to the front with four Staffel aircraft after lunch on a sunny day.

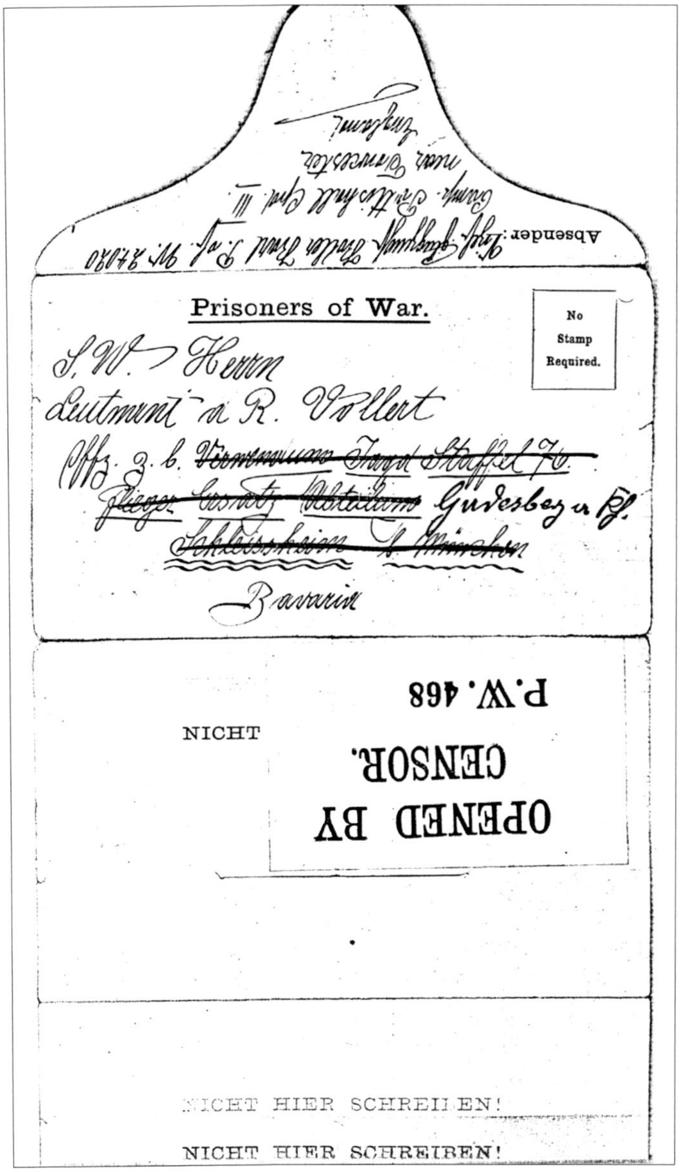

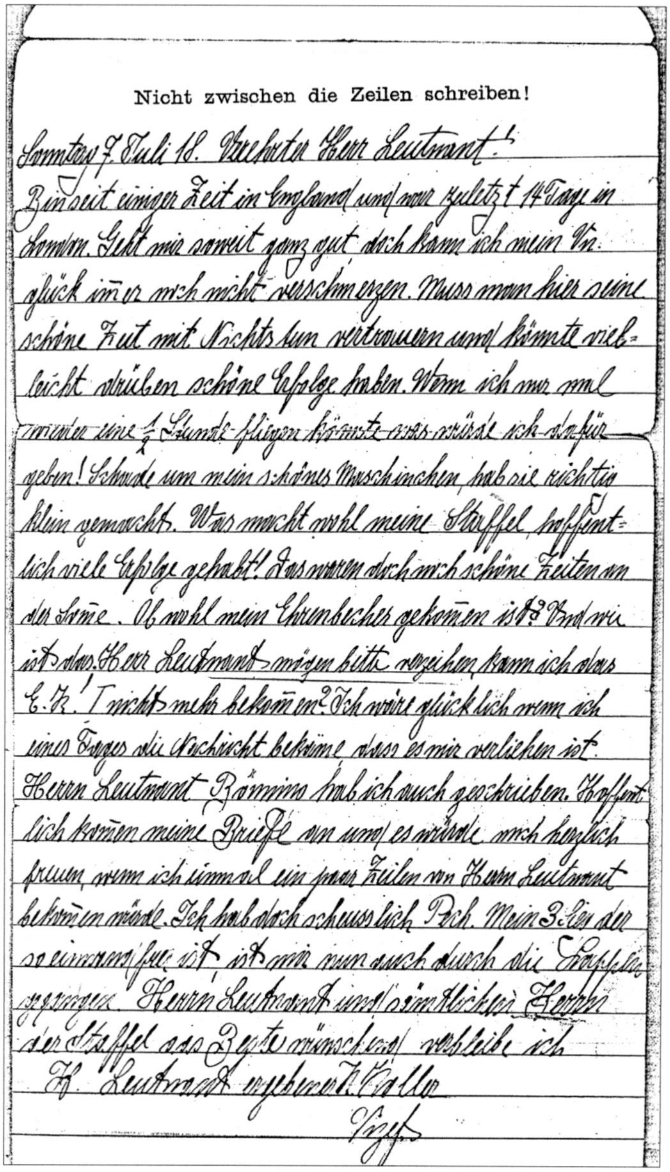
Nicht zwischen die Zeilen schreiben!

Sonntag 7. Juli 18. Werter Herr Leutnant!
Bin seit einiger Zeit in England und war zuletzt 14 Tage in London. Geht mir soweit ganz gut, doch kann ich mein Unglück immer noch nicht verschmerzen. Muss man hier seine schöne Zeit mit Nichtstun vertrauern und könnte vielleicht drüben schöne Erfolge haben. Wenn ich nur mal wieder eine ½ Stunde fliegen könnte, was würde ich dafür geben! Schade um mein schönes Maschinchen, hab sie richtig klein gemacht. Was macht wohl meine Staffel, hoffentlich viele Erfolge gehabt! Das waren doch noch schöne Zeiten an der Somme. Ob wohl mein Ehrenbecher gekommen ist? Und wie ist das, Herr Leutnant mögen bitte verzeihen, kann ich das E.K. I nicht mehr bekommen? Ich wäre glücklich wenn ich eines Tages die Nachricht bekäme dass es mir verliehen ist. Herrn Leutnant Bömming hab ich auch geschrieben. Hoffentlich kommen meine Briefe an und es würde mich herzlich freuen wenn ich einmal ein paar Zeilen von Herrn Leutnant bekommen würde. Ich hab doch scheusslich Pech. Mein 3. Sieg der so einwandfrei ist, ist mir nun auch durch die Lappen gegangen. Herrn Leutnant und sämtlichen Herren der Staffel das Beste wünschend verbleibe ich
Ihr Herrn Leutnant ergebenster K. Koller
Vzfw.

Above: Letter written by Karl Koller on 7 July 1918 to Lt. Clemens Vollert while he was in British captivity.

The five aircraft flew towards the north to respond to reports of Allied flight activity. West of Albert he sighted a formation of British fighters. When the Germans attacked, the British pilots turned around to face them, resulting in a swirling dogfight. He succeeded in separating one of the opposing Sopwith Camels from its formation, pushing him to lower during the dogfight, until he was forced to land in German lines near Contalmaison after receiving numerous hits. The pilot, Lt. W. A. Scott from 70. Squadron RAF, was not injured and was taken into German captivity. He had departed at 12:10 as part of an offensive patrol formation. The victory was credited to Lt. Walter Böning as his 16th aerial victory.

During the fight, Vzfw. Markert pursued another Sopwith Camel across the lines towards Albert. After receiving multiple hits, the British aircraft crashed. The British pilot managed to go down in an emergency landing in Allied territory. As the British aircraft had been seen to go down by German ground observers, Georg Markert received confirmation for his first aerial victory.

In the afternoon around 16:00, the Staffel was involved in aerial combat with multiple Camels from 43. Squadron RAF. Lt. Böning managed to position himself behind one of these aircraft. After receiving multiple hits from his machine guns, the aircraft caught fire and crashed near Bray, in close proximity to its home base. Lt. P. T. Bruce died in the crash; his body was recovered from the charred wreckage of his machine. He had departed on a

Above: The wreckage of Sopwith Camel D1793 flown by Lt. P.T. Bruce, 54 Sqdn., shot down by Lt. Walter Böning on 30 May 1918. It was Walter Böning's 17th and final aerial victory.

fighter patrol flight in the vicinity of Flers - Combles - Albert at 14:40. Walter Böning was credited with this Sopwith Camel as his 17th aerial victory. While he was unaware of the fact on that day, it was also destined to be the last one to be confirmed to him.

The next morning, May 31, 1918, brought about good flying weather. From the front reports arrived noting exceptionally lively enemy flying activity. In response, the Staffel undertook repeated flights to the front that day. During one of these missions, the Staffel turned to the south-west toward Villers-Bretonneux-Harbonnières, thereby crossing the frontlines. Near Hangard, behind enemy lines, Uffz. Josef Diem attacked a French Bréguet 14. At the end of the encounter, after taking numerous hits, the two-seater went down in flames. Although the victory was achieved over Allied territory, a German ground observer must have witnessed the action and provided the required details for the confirmation of the claim. Fortunately for Josef Diem, he was finally granted the confirmation for his 1st officially credited aerial victory. According to Walter Böning, it was highly deserved indeed!

Follwing the completion of their sortie, the Staffel safely landed back at Suzanne airfield. Lt. Carolus Bärenfänger, who was now approaching the very limits of his physical strength, complained of exhaustion, discomfort, and dizziness upon his return. As a consequence, Walter Böning ordered him to stay on the ground. As his own aircraft had received multiple hits in the previous aerial combat and was not immediately airworthy at the time, Walter Böning flew Carolus Bärenfänger's "orphaned" Albatros D Va D.5765/17 on the next mission.

The Staffel turned to the north in order to gain altitude east of Albert toward Montauban in German territory. They were still climbing when a formation of enemy "S.E." aircraft of more than twice their own number approached from the west (the available documents do not offer definitive information as to whether a reference is made to S.E. 5a or Sopwith single seaters, note of the authors), surprisingly attacking them from above. A wild aerial battle ensued which hardly offered any of the pilots a clear shot at the enemy. Suddenly, a ricochet entered Lt. Walter Böning's aircraft, piercing his left calf and fibula. Numb with pain he entered a nosedive and escaped to a lower altitude. He managed to reach his home airfield, which was only about 8 km away, where he made a smooth landing before losing consciousness due to a substantial loss of blood. Clemens Vollert rushed over to the aircraft and pulled him out with the help of several mechanics. They placed him on the back seat of the Staffel car and raced to Field Hospital 97.[21]

Vzfw. Georg Markert, who had been locked in combat with his opponent, was attacked from behind by a second enemy; his aircraft caught fire and crashed near Fricourt at 18:15. He had written a letter to his wife the day before. He started writing the letter in the morning but had to interrupt due to

Right: Vzfw. Georg Markert's award certificate for the "Sieger im Luftkampf" (Victor in Aerial Combat) Honorary Cup, dated June 27, 1918, exactly four weeks after his first and only victory. (R. Kastner)

Großes Hauptquartier, den 27. Juni 1918

An den

Vizefeldwebel Markert

in der Königlich Bayerischen Fliegertruppe.

Vaterländische Freunde der Luftstreitkräfte haben mir eine Geldsumme zur Verfügung gestellt, die ich zur Schaffung eines

Ehrenbechers

für den Sieger im Luftkampfe

verwendet habe.

Es ist mir eine dankbare Freude, Ihnen als Anerkennung für die im siegreichen Gefecht bewiesene Tapferkeit den Becher überweisen zu können.

Der Kommandierende General der Luftstreitkräfte.

Allerhöchst mit Wahrnehmung beauftragt

v. Hoeppner

Generalleutnant.

the front flights. He did not have time to finish the letter until that evening:

"...I began writing this letter in the morning, but I had to interrupt as we had to fly to the front. Now it is 7 o'clock in the evening; I will quickly finish this letter as I will have to go flying again at 8 this evening. Today at 12:30, I shot down my first Englishman in flames. No other news. My Staffelführer has brought down two.

I am sending kisses and am remaining your faithful Georg.

Give my regards to mother and everyone."[22]

While a time is known for the crash of Vzfw. Georg Markert, neither the Staffel KTB nor his personnel file containing the loss report mention the time at which Walter Böning was wounded. According to his personal account, he was injured in the evening during their mission, and did not realize that Vzfw. Markert had been killed. Therefore, it is likely that he was wounded shortly before Georg Markert's crash but during the same mission, at around 18:15.

One can only speculate about the likely opponent of the Staffel in this ill-fated aerial battle. Circumstantial evidence hints at the involvement of Sopwith Camels from 70. Squadron RAF in this encounter. A flight commander of this squadron, Capt. John Todd, M.C., reported bringing down an "Albatros Scout" in flames at 18:00 (British time), south-east of Albert. As Fricourt is located approximately 5 km east of Albert, the claim he reported may well relate to Vzfw. Georg Markert's crash. The German and English time quotations do not match exactly, but no other RAF fighter unit is known to have reported any aerial victories near Albert around that time.

In a letter dated June 2, 1918, addressed to Georg Markert's widow, Unteroffizier Ludwig Artmann reported the events of May 31, which cast more light on the events of that day:

"Dear Frau Markert.

It is my duty to honor the request of a dear comrade. A heavy stroke of fate happened to us on May 31 when your dear husband and my faithful

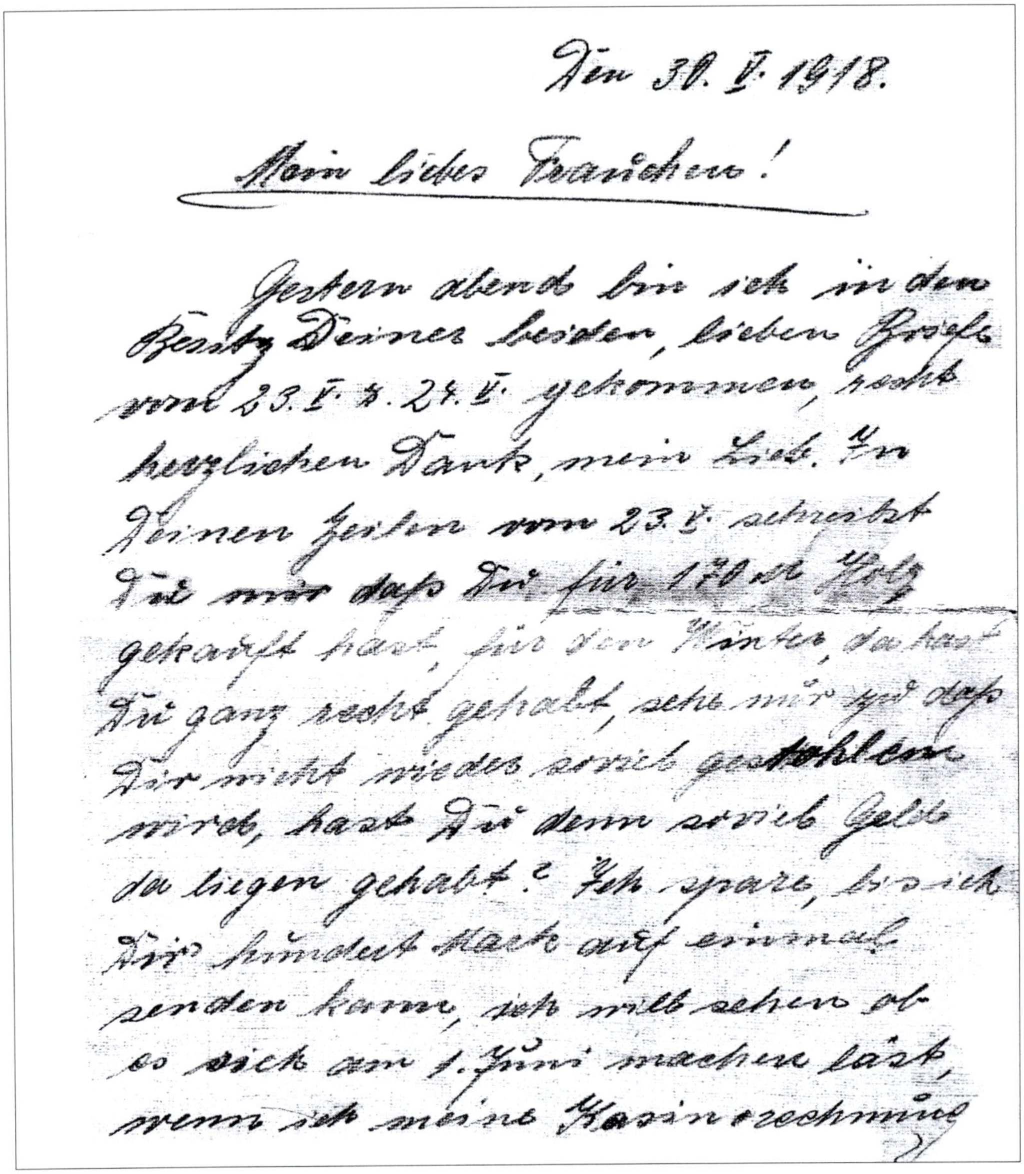
Den 30. V. 1918.

Mein liebes Frauchen!

Gestern abend bin ich in den Besitz Deiner beiden, lieben Briefe vom 23.V. u. 24.V. gekommen, recht herzlichen Dank, mein Lieb. In Deinen Ersten vom 23.V. schreibst Du mir daß Du für 120 M Holz gekauft hast, für den Winter, da hast Du ganz recht gehabt, sehe nur zu daß Dir nicht wieder soviel gestohlen wird, hast Du denn soviel Geld da liegen gehabt? Ich spare, bis ich Dir hundert Mark auf einmal senden kann, ich will sehen ob es sich am 1. Juni machen läßt, wenn ich meine Kasinorechnung

Left & Facing Page: Letter from Georg Markert to his wife, written the day before his fatal crash. (R. Kastner)

roommate paid the ultimate price. On May 31 at 7 o'clock he was severely wounded and succumbed to his wounds on June 1. We lost a dear friend and hero while you lost the dearest person in your life, if I may say so.

I feel for you, valued Frau Markert; I share your pain over the loss of your most prized richness in this world. Please accept my heartfelt condolences. It was his wish that I would contact you, since we shared quarters, after his death if it should come.

I have packed his belongings into a suitcase and will have it brought to you by one of the boys. I will also send you a list of the items contained in the suitcase. He was wearing several of his own clothing items; the other items were taken by the Staffel. Almost all of his personal belongings are in the suitcase, with the exception of the clothes he wore on his last flight.

I will take the liberty of asking you to send me a photograph of him, I would be very much obliged. I will send you a picture of his grave. The suitcase contains a map where I will mark the spot where our faithful friend was buried.

Again, my most heartfelt condolences to you and your family. Signed, your faithful servant,

Flugzeugführer Untoffz.

Ludwig Artmann Bayr. Jagdstaffel 76"[23]

The letter written by Ludwig Altmann proves that Vzfw. Georg Markert was still alive when rescued from the wreck of his Albatros D Va, which had crashed in flames. Unfortunately, he succumbed to his severe injuries the next day. He was laid to rest with military honors on June 2, 1918 at the military cemetery located on a hill between Suzanne and Vaux. His grave was moved after the war; he was

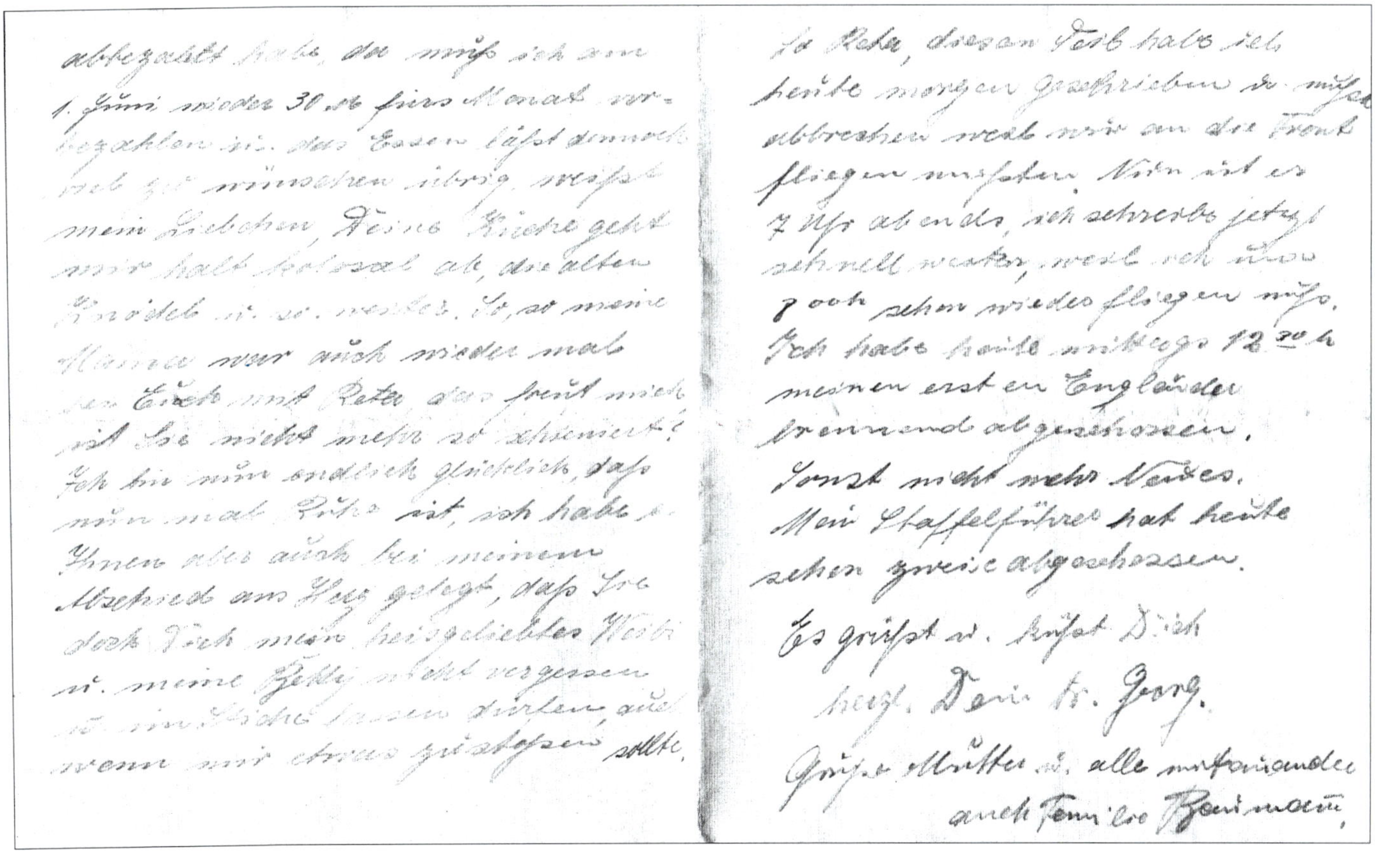
abbezahlt habe, das muß ich am 1. Juni wieder 30 M fürs Monat vorbezahlen u. das Essen läßt dennoch viel zu wünschen übrig, weißt mein Liebchen, Deine Küche geht mir halt kolossal ab, die alten Knödel u. so weiter. So, so meine Mama war auch wieder mal bei Euch mit Peter das freut mich ist Sie nicht mehr so schenierl? Ich bin nun endlich glücklich, daß nun mal Ruhe ist, ich habe Ihnen alles auch bei meinem Abschied ans Herz gelegt, daß Sie doch Dich mein heißgeliebtes Weib u. meine Zelly nicht vergessen u. unglücklich lassen dürfen auch wenn mir etwas zustoßen sollte.

So Resl, diesen Brief habe ich heute morgen geschrieben u. mußte abbrechen weil wir an die Front fliegen mußten. Nun ist es 7 Uhr abends, ich schreibe jetzt schnell weiter, weil ich um 7 30 schon wieder fliegen muß. Ich habe heute mittags 12 30 h meinen ersten Engländer brennend abgeschossen. Sonst nicht mehr Neues. Mein Staffelführer hat heute schon zweie abgeschossen.

Es grüßt u. küßt Dich herzl. Dein tr. Georg.

Grüße Mutter u. alle miteinander auch Familie Baumann.

reburied at the war cemetery at Bray-sur-Somme.

When Bruno Schmäling visited Walter Böning in September of 1976, he had received documents supplied by American aviation historian William R. Puglisi at some earlier time. This material contained information on Jagdstaffel 76. Included here was the indication that, on May 31, 1918 Walter Böning was not only injured but had collided with another Jagdstaffel 76 aircraft in mid-air before he came down. This story has been repeated in several publications; the book "Jasta Pilots" mentions that Walter Böning was involved in a mid-air collision with Georg Marker's aircraft.[24]

Surprisingly, neither Walter Böning's letters nor transcript excerpts of the Staffel war diary contained any information relating to the quoted mid-air collision.

When presented with the documents and asked about a collision in the air, Walter Böning was more than surprised. He shook his head and vehemently denied ever having collided with another Staffel aircraft in the air, especially not on that day. Therefore, this story, in spite of being repeatedly circulated, remains pure fantasy.[25] It appears to have originated by misinterpretations of German documents due to a lack of proper knowledge of the German language.

On June 1, 1918 Walter Böning wrote a letter to his parents from the wounded soldiers collection point:

"I am sure you have received the telegraph with the report of my injury by now. During aerial combat, (a) machine gun bullet ricocheted and badly pierced my left lower thigh, fibula. My shin is lightly injured. The injury left a proper hole which will require me to lie down for several weeks. I can vividly picture aunt Anna saying: "Thank God the boy got a proper blighty wound." I am sure I do not have to mention that I am in pain, but one is used to putting up with that. Two days ago, I shot down two Brits on our side; Now I find myself on the receiving end, but who knows if it´ll be any good. The only bullet that hit the aircraft and me could have had even worse consequences. Today, I will board the hospital train, but I do not yet know where it´s going. I have been fully splinted, and I am having difficulty writing…"[26]

In another letter written on June 5, 1918 he wrote:

"The wound looked pretty bad. The bullet hole had the size of an egg, black and slimy. I have been splinted up to my abdomen and am not allowed to move, or else there will be discomfort, so I will lie still. Please send me newspapers; there are two dear and funny comrades in my room, one of them was also a fighter pilot with one of the Bavarian Jagdstaffeln near me."[27]

Above: The still-smoking wreckage of Vzfw. Georg Markert´s crashed Albatros D Va on 31 May 1918.

Left: Georg Markert's grave at Bray sur Somme. (R. Kastner)

In conversation, Dr. Böning recounted:

"In addition to the adversity caused by my injury, I was constantly tormented by thoughts of what might happen to my men. Besides Diem, there was no one there who could have led them properly, but he alone was not capable of performing miracles. At times I woke drenched in sweat in the middle of the night, thinking one of them had been shot down."[28]

Nearly two weeks later, he reported in a letter written from the field hospital on June 23, 1918:

"An Obersekundar (high school student, 11th school year) showed up, introducing himself as the son of one of my people (Jagdstaffel 76 ground personnel, author's note). He handed me a small package with regards from Mother and Father. After he delivered his well-formulated message, I asked him to take a seat, and we started talking. I really cannot say that I knew these people, and I had not been in particularly close contact with the father

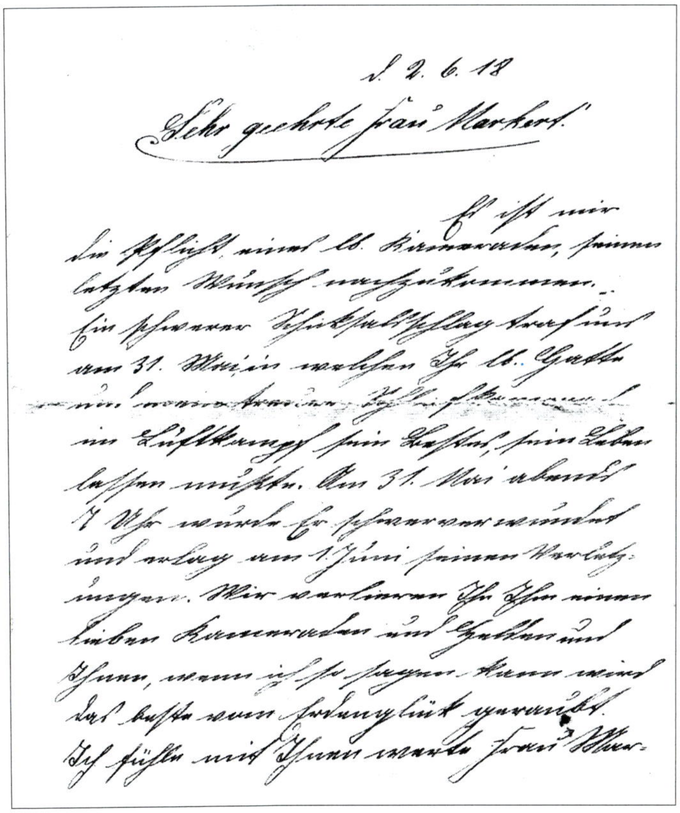

d. 2. 6. 18

Sehr geehrte Frau Markert!

Es ist mir die Pflicht eines lb. Kameraden, seinen letzten Wunsch nachzukommen. Ein schwerer Schicksalsschlag traf Sie am 31. Mai in welchem Ihr lb. Gatte ~~und mein treuer [illegible]~~ im Luftkampf sein Leben lassen mußte. Am 31. Mai abends 7 Uhr wurde er schwer verwundet und erlag am 1. Juni seinen Verletzungen. Wir verlieren ... einen lieben Kameraden und Gatten und Ihnen, wenn ich so sagen kann wird das beste vom Lebensglück geraubt. Ich fühle mit Ihnen werte Frau Mar-

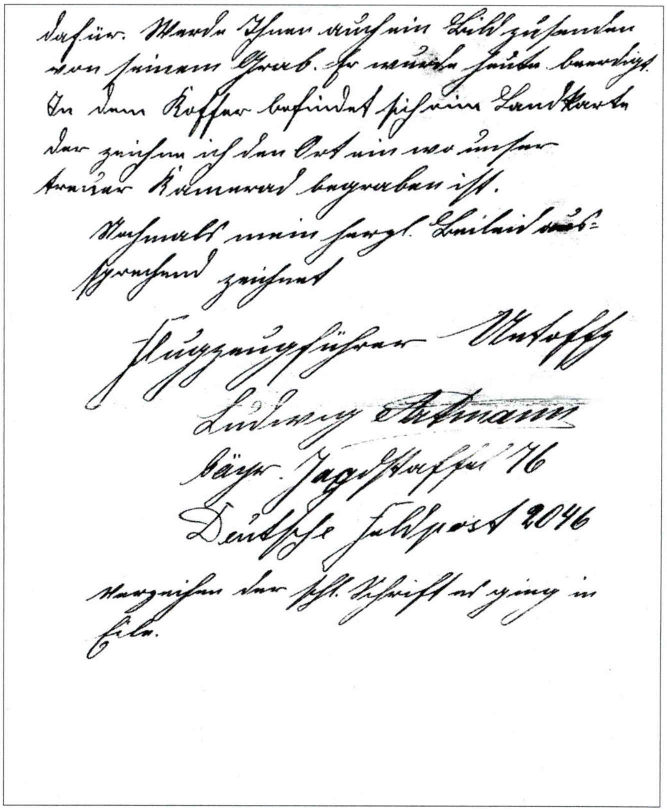

dafür. Werde Ihnen auch ein Bild zusenden von seinem Grab. Er wurde heute beerdigt. In dem Koffer befindet sich eine Landkarte da zeichne ich den Ort ein wo unser treuer Kamerad begraben ist.

Nochmals mein herzl. Beileid aussprechend zeichnet

Flugzeugführer Uffz.
Ludwig Altmann
Bayr. Jagdstaffel 76
Deutsche Feldpost 2046

Verzeihen Sie der schl. Schrift es ging in Eile.

Above & Above Right: Letter from Uffz. Ludwig Altmann addressed to Georg Markert's widow, dated June 2, 1918. Less than six weeks later, Ludwig Altmann fatally crashed due to wing failure. (R. Kastner)

Right: Vzfw. Georg Markert (R. Kastner)

while we were out there. Nonetheless, the son announced that his mother would like to pay me a visit in the coming days. The package contained chocolate and cake, both of the best quality. I was thoroughly amazed and do not yet know how to properly thank them."[29]

He supplemented the letter with the following comments in conversation: *"I always tried hard to maintain a good relationship with all of our ground crew, making sure they were cared for and treated with the deserved respect by their superiors."*[30]

Walter Böning's injury, along with the increasing inferiority of the Albatros D Va, Jagdstaffel 76's fighting power had reached a low point from which it would not recover until the end of the war.

On June 1, 1918 Lt. Max Schick, a largely inexperienced fighter pilot, was appointed deputy Staffelführer. This is an indication of just how short the Bavarian Fliegertruppe was of experienced pilots who were capable of leading a Jagdstaffel. Lt. Max Schick had not even been able to gain the experience required to act as Kettenführer in the air during his short time with the Staffel.

Anerkannte Abschüsse
=====================

des Leutn.d.R. B ö n i n g .

a) als K.E.-Flieger bei Jagdstaffel 19 :

1. Gemäß Kogenluft 103 666 Fl.A v.24.4.17 wird dem Leutnant d.R. B ö n i n g der Abschuß eines einmotorigen Gitterrumpfes am 6.4.17 als erster siegreicher Luftkampf anerkannt.

2. Gemäß Kogenluft 104 414 Fl.Ia v.6.5.17 wird dem Leutnant d.R. B ö n i n g der Abschuß eines Fesselballons am 30.4.17 als 2.siegreicher Luftkampf anerkannt.

3. Gemäß Kogenluft 104 615 Fl.Ia v.8.5.17 wird dem Leutnant d.R. B ö n i n g der Abschuß eines doppelmotorigen Caudron am 29.4.17 als 3.siegreicher Luftkampf anerkannt.

4. Gemäß Kogenluft 109 543 Fl.IIIc v.25.8.17 wird dem Leutnant d.R. B ö n i n g der Abschuß eines Rumpf-Caudron am 17.8.17 als 4.siegreicher Luftkampf anerkannt.

5.u.6. Gemäß Kogenluft 111 196 Fl.IIIc wird dem Leutnant d.R. B ö n i n g der Abschuß von zwei Nieuport-Einsitzern am 23.9.17 als 5. und 6. siegreicher Luftkampf anerkannt.

b) als Führer der Bayr.Jagdstaffel 76 :

7. Gemäß Kogenluft 114 818 Fl.IIIc v.18.12.17 wird dem Leutnant d.R. B ö n i n g der Abschuß eines Spad's am 9.12.17 als 7.siegreicher Luftkampf anerkannt.

8. Gemäß Kogenluft 117 614 Fl.IIIc v.22.2.18 wird dem Leutnant d.R. B ö n i n g der Abschuß eines Paul Schmitt am 11.2.18 als 8.siegreicher Luftkampf anerkannt.

9. Gemäß Kogenluft 117 416 Fl.IIIc v.23.2.18 wird dem Leutnant d.R. B ö n i n g der Abschuß eines Spad-Flugzeuges am 4.2.18 als 9.siegreicher Luftkampf anerkannt.

10. 11. Gemäß Kogenluft 118 765 Fl.IIIc v.22.3.18 wird dem Leutnant d.R. B ö n i n g der Abschuß zweier Spad-Einsitzer am 6.3.18 als 10. und 11. siegreicher Luftkampf anerkannt.

12. 13. Gemäß Kogenluft 119 342 Fl.IIIc v.1.4.18 wird dem Leutnant d.R. B ö n i n g der Abschuß zweier Sopwith am 18.3.18 als 12. u. 13. siegreicher Luftkampf anerkannt.

14. Gemäß Kogenluft 121 391 Fl.IIIc v.23.4.18 wird dem Leutnant d.R. B ö n i n g der Abschuß eines Sopwith-Einsitzers am 6.4.18 als 14.siegreicher Luftkampf anerkannt.

Left: List compiled by Lt. Walter Böning's of the Kogenluft confirmations of his victories.

Nevertheless, he managed to shoot down a "British single seater fighter Spad" across enemy lines near Millencourt (west of Albert) on June 7, 1918. As the crash had been observed from the German side, he received credit for his 1st aerial victory. So far, it has not been possible to determine which type of aircraft was actually shot down on this occasion. The only English squadron flying Spad XIII aircraft on that particular frontline sector (23. Squadron RAF) had already given up their Spads in exchange for the Sopwith Dolphin several months earlier.

On June 17, 1918 Leutnant der Reserve Maximilian Veith arrived at the Staffel, followed by the arrival of Leutnant der Landwehr Karl Dumler, both had graduated from Jagdstaffelschule I.

Maximilian Veith was born in Lindau am Bodensee on October 5, 1895. At the beginning of the war, he deployed with the 3rd Bavarian Infanterie-Regiment as Fahnenjunker on August 17, 1914. During the first years of the war, his Regiment largely fought on the Eastern front where he sustained injuries to his right shoulder on July

15, 1915. He had been promoted to Leutnant on June 2, 1915. Upon his recovery, he returned to his Regiment. On December 18, 1917 he was commanded to Bavarian Flieger-Ersatz-Abteilug II in Fürth for three months. On February 4, 1918, he began pilot training at the local Militär-Fliegerschule 3. His transfer to Armee-Flugpark 6 followed on June 9, 1918; he attended and completed fighter pilot single seater training at Jagdstaffelschule I, followed by his transfer to Jagdstaffel 76.[31]

Karl Dumler was Swabian, from Augsburg. At the age of 32, he was significantly older than all the other Staffel pilots. He was born on February 28, 1886, and as early as 1906 he had already completed his two-year volunteer service with the 4th Bavarian Feldartillerie-Regiment. He was working as a merchant in his hometown of Augsburg. At the beginning of the war, Karl Dumler re-entered service as Unteroffizier der Reserve with Bavarian Reserve-Feldartillerie-Regiment 1. During a battle near Arras on June 16, 1915 Karl Dumler sustained grave injuries to his right knee joint, right upper thigh, and left upper arm caused by shrapnel. A few days after his promotion to Leutnant der Landwehr I on October 8, 1917, he transferred to the Fliegertruppe. On October 16, 1917 he was commanded to begin pilot training with Bavarian FEA 1 in Schleißheim. He began practical flight training at Militärfliegerschule 1 on November 26, 1917. On February 9, 1918, after completion of 39 training flights, he went aloft on his first solo flight.

Dumler's assessment issued by Oberleutnant Oskar Falkner von Sonnenburg, head of the flight school, attested his good airmanship as well as his sound knowledge of the aircraft and its engine. He displayed consistent enthusiasm and diligence with the strictest conception of duty. His military essence and conduct while off-duty were always impeccable, making him well suited to serve as a fighter pilot. On June 6, 1918 he was transferred to AFP 2 and commanded to Jagdstaffelschule I. His training was concluded by his transfer to Jagdstaffel 76.[32]

After the war had ended, Oblt. Amandus Rostock, the last Staffelführer assessed Dumler's performance:

"Leutnant Dumler displayed a lot of solicitude and conscientiousness during his time with the Staffel. His conception of duty must be described as being exemplary. No matter how big or small the task Dumler was assigned, he was reliable and executed his duties with diligence, even under the most difficult of circumstances. He proved himself particularly during the last weeks removal of the Staffel (from the front of the 1st German Army, note of the authors) and the equipment was recovered. He was a very good aviator. His flying skills were well polished. On November 1, he achieved his first confirmed aerial victory.

To his comrades, Dumler displayed the utmost courtesy and devotion which made him quite popular among them. While off-duty, he never forgot that he was an officer. His appearance was calm and confident. His manners are good."[33]

In exchange for the two new arrivals, Lt. Karolus Bärenfänger who was completely worn-out, was eventually transferred to Bavarian Flieger-Ersatz-Abteilung I in Schleißheim on June 21, 1918.

His Staffelführer, Walter Böning had issued the following assessment of him on May 27, 1918:

"Lt. Bärenfänger was one of the first pilots in the newly formed Staffel. Due to his amiable and comradely behavior he was well-liked by everyone. His duties in the Staffel included the position of technical and truck officer, which he executed to my full satisfaction. As a pilot, he always met my expectations. He was unsuited to serve as Kettenführer as he was lacking eagerness to attack. I consider him a suitable candidate for a post as office officer, aerial protection officer, or similar, in spite of his used nerves. As he had not undergone any formal military training (commissioned officer's training, author's note) his conduct in service was not always unobjectionable. His funds are in order.

Böning, Lt.d.R. and Staffelführer"[34]

As an amendment to this assessment, Walter Böning remarked:

"He came from the non-commissioned rank and as an officer was always very insecure in dealing with subordinates, so he didn't dare give orders or monitor compliance. In his private life, he was a likeable and helpful person with a high sense of duty."[35]

One day later, Uffz. Josef Diem was finally awarded the deserved EK I (Iron Cross 1st Class).

Max Schick's tenure as deputy Staffelführer was of short duration; the following day, word reached the Staffel that Leutnant der Landwehr Ludwig Schmid from Bavarian Jagdstaffel 34 had been designated the new Staffelführer.

Ludwig Schmid was born on January 16, 1894 in Lauf near Hersbruck in Central Franconia. He joined the Kadettenkorps of the 8th Bavarian Feldartillerie-Regiment as Fähnrich (sergeant) on August 13, 1914. On June 2, 1915 he was promoted to Leutnant, retroactively as of September 19, 1914. At the end of 1915, he applied for observer training. On January 8, 1916 he was commanded to Bavarian Flieger-Ersatz-Abteilung 1 in Schleißheim. Upon completion of his training, he attended Artillerie-Beobachter-Kurs in Jüterbog on February 15, 1916. He was transferred to Bavarian Feldflieger-Abteilung 2 on April 20, 1916 via Armee-Flug-Park 6a. After about nine months and many missions flown in the observer seat, he volunteered for pilot training. Via Armee-Flugpark 6a, he initially transferred to Bavarian Flieger-Ersatz-Abteilung 1 on April 7, 1917 before attending pilot training at Militär-Fliegerschule 3 in Fürth. Upon completion of his training, he was commanded to Jagstaffelschule II in Nivelles on November 27, 1917. One week later, he arrived at Jagdstaffel 34. His new Staffelführer, Oblt. Robert Greim, had a reputation for being a very good Staffelführer who put a lot of emphasis on the training of his pilots.[36] On March 23, 1918 Lt. Ludwig Schmid achieved his first aerial victory over a De Havilland D.H. 4, and one day later over an S.E. 5a. Both kills were later confirmed by Kogenluft. With his addition, the Staffel received an experienced fighter pilot as Staffelführer.[37]

On June 23, 1918, the day when Lt. Schmid was designated Staffelführer, Max Schick departed on a solo front flight at approximately 20:15 on Albatros D Va D.5767/17 (the former aircraft of Lt. Bärenfänger) under cloudy skies. He latched onto a Jagdstaffel 46 Kette cruising along the front at an altitude of approximately 3700 m. The pilots spotted the lone Jagdstaffel 76 aircraft that followed them toward Albert, about 700 m above their level, before losing sight of it.

More than 85 years later, the remains of his aircraft were discovered about 40 km south of Albert in a swamp. What happened?

Comparing the area where Max Schick was last seen by the Jagdstaffel 46 pilots with the site of the crash, he must have first flown across the front in the vicinity of Albert, heading westbound before turning south above Allied lines. The route of flight, time of day, along with the fact that he was flying alone suggest that he intended to attack an Allied captive balloon.

Heinrich Arntzen, former Jagdstaffel 15 Jagdflieger and Jagdstaffel 50 Staffelführer, who himself had shot down four captive balloons, explained the typical tactic of attacking captive balloons: the pilot would cross the front at high altitude and at a great distance from the "gasbag". Then, the pilot would continue to fly along the enemy front at high altitude, using the clouds as cover whenever possible. Once in the area of the captive balloon, the engine was throttled down, or ignition was switched off altogether, and the pilot carefully descended upon his target, making the best possible use of available cloud cover. It was crucial that the fighter pilot was discovered as late as possible, so he could attack the captive balloon while approaching from his own lines and set it on fire. Once this was accomplished, the fighter pilot had to escape as quickly as possible to avoid being attacked himself; all available anti-aircraft guns would upon fire on him once his presence was noted. The pilot then would take advantage of the tailwinds in order to make it back across his own lines, twisting and curving to avoid the hail of fire.[38]

Attacking an enemy observation balloon was one of the most dangerous missions for fighter pilots during WWI. These objects were of great military significance as they had the ability to spy far into enemy rear areas, capture photographs, and direct their own artillery fire on concealed enemy positions. When a captive balloon was shot down it usually took about a day or two before a new one was brought to the same location, filled up with gas, and moved into position. During that period of time, artillery gunners lacked well-guided direction and were unable to receive any information regarding enemy position. For this reason, the balloons were well protected by anti-aircraft artillery and machine gun positions. On the French side, even fighter aircraft were dispatched in addition in order to provide better protection. Shooting down this well-protected target was of greater military value than shooting down an aeroplane, earning the pilot a great deal of respect.

Captive balloon specialists such as Oberleutnant Fritz Röth (Jagdstaffel 23 and Staffelführer Jagdstaffel 16) as well as Leutnant Heinrich Gontermann (Jagdstaffel 5 and Staffelführer Jagdstaffel 15) gathered all available information on the exact position of the balloon along with terrain information, ground defense, and presence of enemy fighter aircraft. Regularly they drove to the front themselves in order to observe the balloons they were planning to attack before planning their mission general staff style, awaiting ideal weather conditions for several days if necessary. According to Heinrich Arntzen, Heinrich Gontermann trained his Jagdstaffel 15 pilots for attacks on captive balloons. Such attacks were to be conducted by pilots on their own accounts, but rather after careful planning, and with express order to conduct such an attack.[39]

Whether Max Schick's intention to attack a captive balloon was an ad hoc plan, and whether he overestimated his own abilities or was hoping to prove his talents as a potential Staffel leader are only speculative. As a newcomer with little fighter pilot or front flying experience and having only little knowledge of attacks on captive balloons, his chances for a successful execution of this plan were slim from the beginning.

In the Contoire area north of the town of Boussicourt, approximately 12 km west of Roye, he attacked a captive balloon. His aircraft was hit by anti-aircraft groundfire, and subsequently crashed.

Above: Lt. Maximilian Veith arrived on the 17 June 1918 and remained with the unit until the end of the war. (R. Kastner)

Above: Lt. Ludwig Schmid, the new Staffelführer of Jasta 76, joined the unit coming from Bavarian Jagdstaffel 34. He had served as a pilot with this unit for seven months.

Lt. Max Schick managed to free himself from the falling aircraft by taking to his parachute. After his landing, he was taken captive by French soldiers. The aircraft crashed into a remote swamp area near a forest in the vicinity of Boussicourt.

Anti-aircraft artillery stationed in the area of the French 1st Army reported the kill of a German aircraft that day whose pilot had been taken into captivity. It should be noted that Max Schick was released from French captivity in February of 1920 and returned to Germany.

On July 5, 1918 Gefreite Josef Reinhard and Johann Brunner transferred to the Staffel, joining the unit from Jagdstaffelschule I.

Josef Reinhard was born on March 15, 1889 in Grünmorsbach near Aschaffenburg. On October 14, 1910 he joined Lothringisches Infanterie-Regiment Nr. 144, from which he was discharged to the Reserves two years later. He was an electrical engineer in civilian life, working in Würzburg. When the war began, he joined the 4th Bavarian Reserve-Infanterie-Regiment as a war volunteer. On July 30, 1916 he was transferred to Armee-Flugpark 5 in Thonne-le-Près near Montmédy, likely due to his civilian work skills. On November 1, 1916 he transferred to Armee-Flug-Park 4 in Gent. His request of becoming a pilot was granted. On July 18, 1917 he was first commanded to Bavarian Flieger-Ersatz-Abteilung 2 in Fürth; three weeks later he joined Militär-Fliegerschule 3 at Fürth. Upon successful completion of his training he was transferred to AFP 2. He was then commanded to Jagdstaffelschule I in Valenciennes on June 24, 1918 in order to familiarize himself with fighters. Thereafter, he was transferred to Jagdstaffel 76.[40]

Johann Brunner was born in Bottenbach near Pirmasens in the Palatinate region on February 15, 1897. He signed up to be a war volunteer and joined the Bavarian 14th Infanterie-Regiment on August 8, 1914. Upon approval of his request to join the Fliegertruppe, he was commanded to Bavarian Flieger-Ersatz-Abteilung 1 in Schleißheim on September 1, 1917, and later transferred to

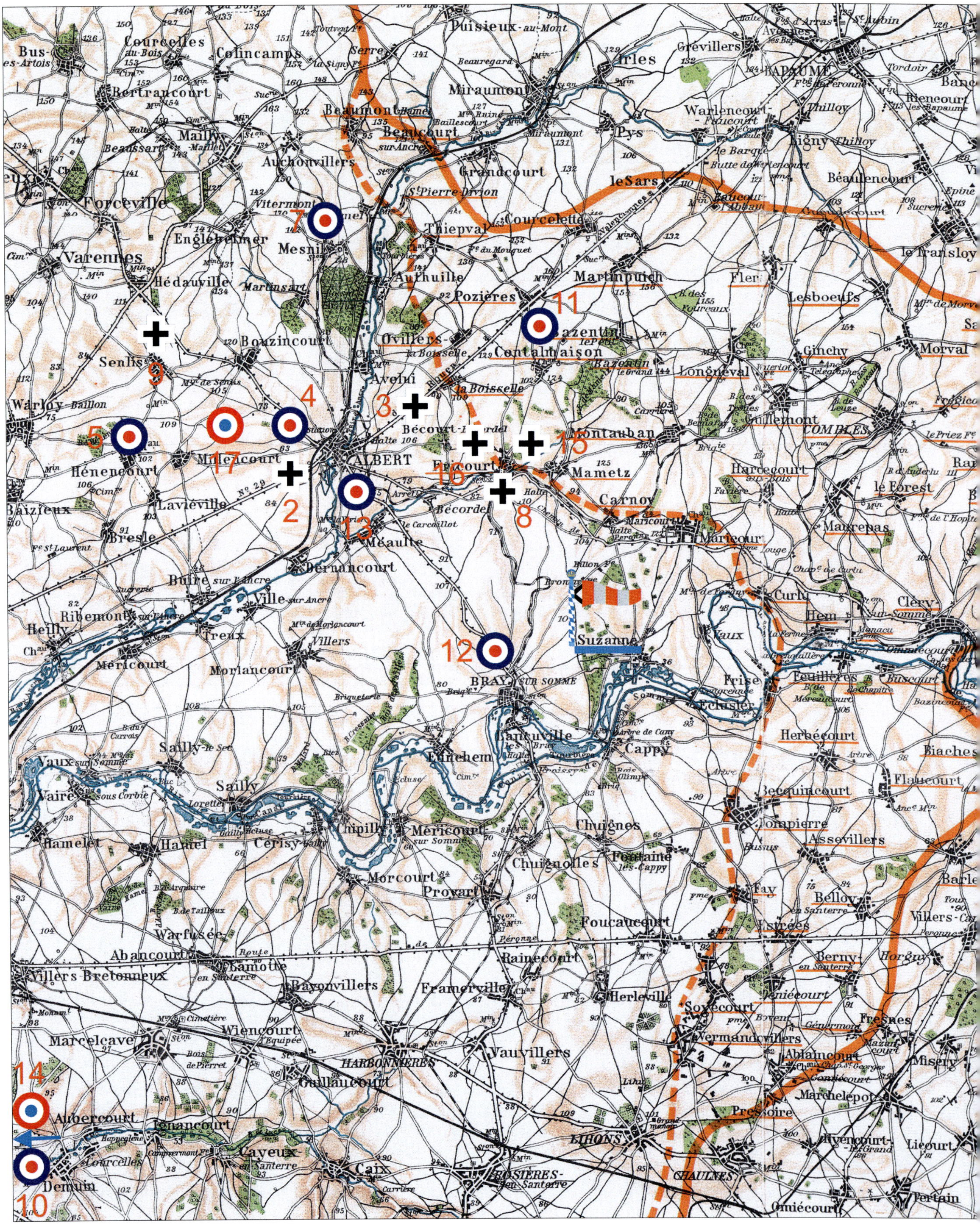
Bus-es-Artois
Courcelles du Bois
Colincamps
Serre
Puisieux-au-Mont
Irles
Grévillers
BAPAUME
Bertrancourt
Beaumont Hamel
Miraumont
Warlencourt
Mailly-Maillet
Beaussart
Auchonvillers
Beaucourt sur Ancre
Grandcourt
Pys
Forceville
Mesnil
St Pierre Divion
Thiepval
Courcelette
le Sars
Beaulencourt
Engelbelmer
Varennes
Hédauville
Martinsart
Authuille
Pozières
Martinpuich
Flers
Lesboeufs
le Transloy
Senlis
Bouzincourt
Ovillers-la Boisselle
Contalmaison
Bazentin
Longueval
Ginchy
Morval
Warloy-Baillon
Aveluy
la Boisselle
Bécourt
Montauban
Guillemont
COMBLES
Hénencourt
Millencourt
ALBERT
Fricourt
Mametz
Hardecourt aux Bois
le Forest
Baizieux
Laviéville
Bécordel
Carnoy
Maricourt
Maurepas
Bresle
Méaulte
Dernancourt
Buire sur l'Ancre
Ville-sur Ancre
Curlu
Hem
Cléry-sur-Somme
Heilly
Ribemont sur l'Ancre
Treux
Méricourt
Morlancourt
Suzanne
Vaux
BRAY SUR SOMME
Frise
Eclusier
Feuillères
Buscourt
Herbécourt
Biaches
Sailly-le-Sec
Sailly
Vaux sur Somme
Vaire sous Corbie
Etinehem
Cappy
Flaucourt
Becquincourt
Dompierre
Assevillers
Hamelet
Hamel
Cérisy-Gailly
Chipilly
Méricourt sur Somme
Chuignes
Chuignolles
Fontaine les Cappy
Morcourt
Proyart
Fay
Belloy en Santerre
Barleux
Villers-Carbonnel
Warfusée
Abancourt
Foucaucourt
Estrées
Berny en Santerre
Villers Bretonneux
Lamotte en Santerre
Bayonvillers
Framerville
Rainecourt
Herleville
Soyécourt
Deniécourt
Fresnes
Marcelcave
Wiencourt
Vermandovillers
Ablaincourt
Misery
HARBONNIÈRES
Vauvillers
Guillaucourt
Marchélepot
Aubercourt
Ignaucourt
Cayeux en Santerre
Caix
Pressoire
Hyencourt-le-Grand
Licourt
Courcelles
Demuin
ROSIÈRES-en-Santerre
LIHONS
CHAULNES
Omiécourt
Pertain
1
2
3
4
5
7
8
9
10
11
12
13
14
15
16
17

Militär-Fliegerschule 2. After completing his training, Brunner was commanded to Bavarian Fliegerabteilung (A) 291 on April 7, 1918. His stay with FA (A) 291 lasted nearly two months before he returned to Armee-Flugpark 6. From there, he transferred to Jagdstaffelschule I and thereafter to Jagdstaffel 76.[41]

Three days later, Leutnant der Reserve Walter Curtius arrived at the Staffel with orders backdated to June 27, 1918. He was born in Duisburg on September 17, 1897. Before the war, he studied philosophy in Munich, deploying with the Ulanen Regiment Nr. 5 of Westphalia as a volunteer when the war began. In 1916, he transferred to Ulanen-Regiment von Katzler (Silesian) Nr. 22, where he was promoted to Leutnant der Reserve on May 25, 1917. At that time, the era of the proud cavalrymen had long passed, so he signed up with the Fliegertruppe. On November 9, 1917, he was commanded to Fürth for 8 weeks in order to attend pilot training with Bavarian Flieger-Ersatz-Abteilung 2. His further training was completed at Militär-Fliegerschule 7 in Germersheim from January 1918. After graduating from pilot training, he transferred to Armee-Flugpark 2 on June 13, 1918 and on to Jagdstaffelschule 1 where he completed training in single seater fighter aircraft and was subsequently transferred to Jagdstaffel 76.[42] With Lt. Curtius, yet another novice fighter pilot joined the unit.

After the war, Oberleutnant Rostock, the last Staffelführer supported Lt. d. R. Curtius' application for the EK I, as detailed in a letter dated July 12, 1919:

"Leutnant d.R. Curtius was discharged from Flugplatzkommandantur Sonthofen to Bez.Kdo. Munich on December 1, 1918. For the last three months or so he has been a member of Husaren-Regiment 11, presently stationed in Berlin. From June 26, 1918 until the unit was disbanded, Lt. Curtius was a member of Jagdstaffel 76. During his time with the Staffel, he was an exceptionally capable soldier and a skillful aviator. His sense of duty, conception of duty, and his eagerness to fly were exemplary. He displayed a steadfast flight discipline while flying in formation, as well as a superb fighting spirit and courage during his 70 - 80 front flights on the Somme river, in the Champagne region, and in Upper Alsace."[43]

The following day, Jagdstaffel 76 was transferred away from the area of operation of the German 2nd Army. Aircraft, equipment, and personnel were loaded onto trains, and transferred to German 1st Army on the Champagne front.

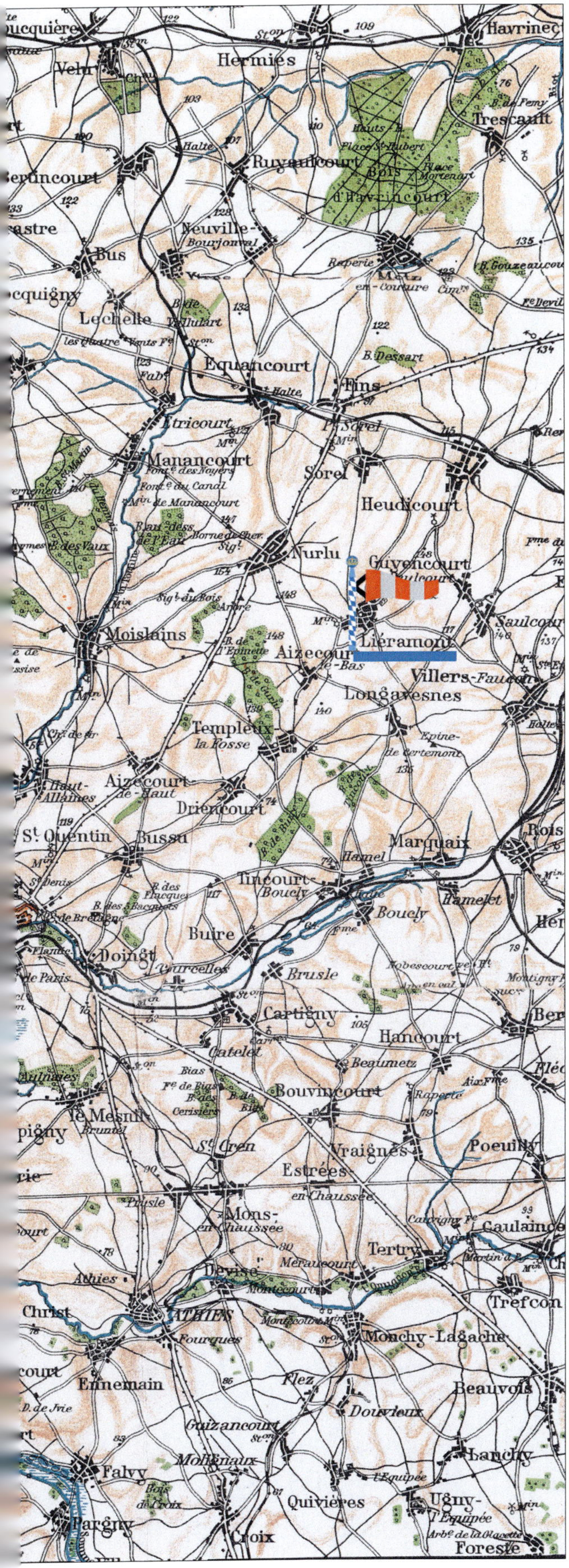

Left: The red line shows the front line before the German offensive in 1918.

German 2nd Army, Airfields Liéramont & Suzanne

Locations & Events

#		Date	Pilot	Location	Footnote
1.		03.04.18	Lt. Walter Böning	?	Sopwith Camel
2.		11.04.18	Lt. Richard Emmerich	4 km west of Albert	Killed in action
3.		11.04.18	Lt. Gottfried Stemmler	North-east of Albert	Killed in action
4.		11.04.18	Lt. Walter Böning	West of Albert	S.E. 5a
5.		11.04.18	Uffz. Josef Diem	Henencourt	R.E. 8 (Forced to land)
6.		02.05.18	Lt. Moritz Weig	Between Dompierre & Godenvillers	Prisoner of war
7.		17.05.18	Vzfw. Karl Koller	Mesnil	D.H. 4
8.		23.05.18	Oblt. Walter Angermund	Fricourt	Forced to land
9.		25.05.18	Vzfw. Karl Koller	Senlis	Prisoner of war
10.		28.05.18	Uffz. Josef Diem	Hangard	S.E. 5a (Forced to land)
11.		30.05.18	Lt. Walter Böning	Contalmaison	Sopwith Camel
12.		30.05.18	Lt. Walter Böning	Bray sur Somme	Sopwith Camel
13.		30.05.18	Vzfw. Georg Markert	Albert	Sopwith Camel
14.		31.05.18	Uffz. Josef Diem	Hangard	Breguet 14
15.		31.05.18	Lt. Walter Böning	Fricourt	Wounded in action
16.		31.05.18	Vzfw. Georg Markert	Fricourt	Killed in action
17.		07.06.18	Lt. Max Schick	Millencourt	Spad
18.		23.06.18	Lt. Max Schick	Boussicourt	Jumped by parachute Prisoner of war

Right: The flight route of Lt. Max Schick on 23 June 1918.

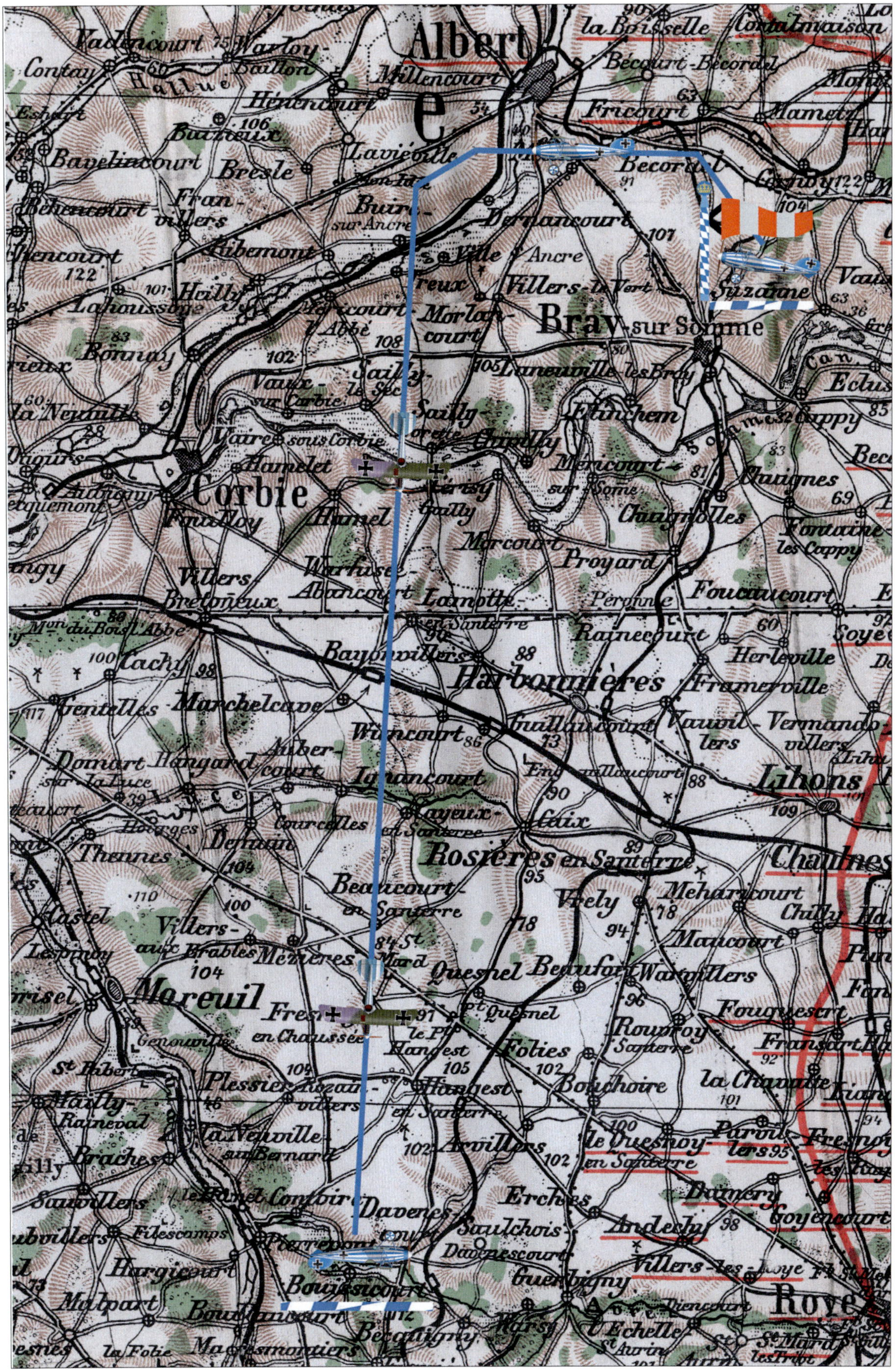

5 Offensive Battle on the Marne River and the Champagne Region

1st German Army, Bignicourt Airfield, July 9, 1918 – August 18, 1918

Left: Prior to joining the Fliegertruppe, Uffz. Josef Bauer had seen service with the German "Alpencorps" (Mountain Troops). His cap was adorned with an "Edelweiß" badge, which was the Troop's badge of honour, given to them by their Austrian allies.

On July 11, 1918 the Staffel arrived in the area of the German 1st Army. The operational area of the German 1st Army in the Champagne region extended from Goeux in the west to west-southwest of Reims and then east, abeam Aubérive.

The airfield was located on flat terrain near the village of Bignicourt, approximately 12 km south of Rethel, and about 32 km north-east of Reims. On July 6, 1918, Royal Prussian Jagdstaffel 63 had resumed operations from there, under the Command of Lt. Hermann Leptien, a native of the city of Kiel. He welcomed the Bavarians with his calm, nordic and cordial personality, instructing his ground personnel to assist their southern German countrymen with all work related to setting up their Staffel on the airfield.

Above: Uffz. Josef Bauer pictured here during his pilot training at Militär-Fliegerschule (aviation-training-camp) 4 in Lager Lechfeld.

Jagdstaffel 76 was combined with Jagdstaffeln 22, 63, and 69 to form Jagdgruppe C, led by the experienced Lt. Alfred Lenz, Staffelführer of Royal Saxon Jagdstaffel 22. Lt. Lenz was a native of Munich but had joined the Saxon Army and was therefore a Saxon officer. His Staffel was stationed at the nearby Alicourt airfield, located approximately 6 km to the west. As during their previous operations in the Alsace, their primary opponents in the Champagne region were French air units; the local units in the Champagne were strengthened by small RAF contingents (predominantly in the western sector of operations), as well as flying units of the American Expeditionary Forces.

Even during the inter-war years, no war diary for Jagdstaffel 76 covering the timespan from June 17, 1918 until the end of the war was held by the Reich Archives in Potsdam. Thus, the information covering this timespan was extracted from various sources and is therefore undoubtedly incomplete.

The deployment in the new offensive sector began with a crash. One day after their arrival, Uffz. Artmann departed on a training flight around 21:00. Shortly after take-off at an altitude of approximately 400 m, one of his wings suffered structural failure. The aircraft crashed 2 km to the north of the airfield. Ludwig Artmann died in the crash; his body was recovered from the wreckage of his Albatros. He was buried on the cemetery of honor in Juniville two days later.

On July 15, 1918, what was to be the last German offensive of the war was launched under code name "Marneschutz-Reims". It involved 23 Divisions of the German 1st and 3rd Armies against the French 4th Army east of Reims, as well as 17 divisions of the German 7th Army against the French 6th Army, west of the city and on the Marne River. The 7th Army was to cross the Marne River near Dormans and Chateau Thierry in the western offensive sector; thereafter, proceeding toward Epernay at a width of 50 km on both sides of the Marne. The attack quickly turned into a failure as Allied Forces had expected their arrival. After some initial minor German ground gains, their opponents were able to force them back to their starting positions following a counterattack.

The High Command of the German 1st Army had probably issued orders to attack enemy observation balloons in an effort to support the German attack. For this reason, Staffelführer Lt. Ludwig Schmid departed on a mission to attack a captive balloon on July 16, 1918. He was unsuccessful however, as he was subjected to heavy anti-aircraft fire, suffering a wound in his right upper thigh. He escaped across the lines to the German side where he made a successful emergency landing. German soldiers recovered him from his aircraft and brought him to the closest field hospital.

On the same day, Lt. Edmund Neu was attacked by Spad single seaters and sustained light injuries. He also managed to escape across German lines, and he safely landed back at his home field of Bignicourt. His injuries were initially tended to at the Staffel sick bay before he was sent home on sick leave on July 21, 1918. He returned to the Staffel on August 12, 1918.

The available documents do not contain any statements regarding deputy leadership of the Staffel in the absence of Lt. Ludwig Schmid due to his injury. It is likely that Offizier z.b.V. Lt. Clemens Vollert was nominally in charge of official Staffel business. It is no longer possible to determine whether one of the Officer pilots (all of them newcomers) was "officially" tasked with Staffel leadership. The authors do not consider this to be a likely scenario. On June 22, 1918, Lt. Hermann Obermeier returned to the Staffel. He had been on sick leave since being injured during a forced landing on May 11, 1918.

One day later, Unteroffizier Josef Bauer and Gefreiter Julius Geyer arrived at the Staffel from Jagdstaffelschule II.

Josef Bauer was born in Aukofen near Regensburg in the Upper Palatinate region on October 3, 1897. His father was a coachman and lived in Regensburg with his wife, Karoline. His personnel file records that he worked as a professional jockey in civilian life.

Further explaining on this, his daughter Luise Wildmann related in conversation:

"As a young man, my father worked as a professional jockey and regularly took part in horse races. He truly was devoted to horses; he literally grew up with them along with the job of his father. He told me that he would have loved to join the cavallery but was not accepted for service there. So, he volunteered for the flying services. When he met my mother (following the end of the war; author´s remark), he was working again as a jockey and trained horses. As a consequence, following their wedding, they travelled from one horse race to the other. For both, it was a beautiful time, free of worries."[1]

He entered the war as a member of the Bavarian 3rd Infanterie-Regiment on August 2, 1914. On June 14, he transferred to the Infanterie-Leib Regiment. Following nearly three years of service as an infantryman, he was looking for a new assignment, therefore he volunteered to join the Fliegertruppe. On November 27, 1917 his training began at

№ 856

Josef Bauer's Certificate of Possession for the Badge as Military Aircraft Pilot.

Besitz-Zeugnis.

Das von Seiner Majestät dem König Ludwig III. von Bayern, Allergnädigst genehmigte

„Abzeichen für Militär-Flugzeug-Führer"

ist dem

Unteroffizier Bauer Josef

der bayr. Jagdstaffel 76.

ausgehändigt worden.

Hierüber wird mit Allerhöchster Ermächtigung dieser Beglaubigungsschein ausgefertigt.

München, Dezember 1917.

B. Insp. d. Militär-Luftfahr-Wesens

~~Inspektion des Ingenieur-Korps~~

S.

Major u. stellv. Insp. d. M. L. W.

Bemerkungen.

1. Tragen des Abzeichens: V. Bl. No. 22/1916 S. 450 Ziff. 1.
2. Das Abzeichen ist solange zu tragen, als der Beliehene in der Liste der Militär-Flugzeugführer geführt wird.
3. Inhaber des Abzeichens, die von der Liste gestrichen sind, haben es mit dem Besitzzeugnis grundsätzlich sogleich nach Empfang der Mitteilung hierüber an ihre nächstvorgesetzte Dienststelle abzuliefern.

Left: Portrait photo of Gefr. Julius Geyer wearing his flying cap and goggles. He joined Jagdstaffel 76 on 23 July 1918 and remained with the Staffel until the end of the war.

Facing Page: Assessment of Julius Geyer written by Oblt. Geier, the commander of Fliegerschule 4.

Jul. Geyer
Marschallstr. 33

Beurteilung

des Flugschülers Uffz. Julius Geyer

mit 17. Bedingung

1. Kam zur Schule 1 am 22. 1. 18 als Flugschüler
2. a) 1. Schulflug am 9. 4. 18 Schulflüge 53
   b) 1. Alleinflug am 27. 4. 18 c) 17. Bedingung am 6. 6. 18
3. Fliegerische Veranlagung: sehr gut große Gewandtheit u. Geistesgegenwart
4. Motorenkenntnisse: sehr gut
5. Eifer und Fleiß: sehr groß
6. Geflogene Typen: C I, C III, C V, C VII, L.V.G., D.F.W., A.E.G., Rumpl. C I, D III
7. Beschädigte Maschinen: 1 leicht (ohne Verschulden)
8. Dienstauffassung: sehr gut
9. Militärisches Wesen: energisch & stramm
10. Außerdienstliches Verhalten: sehr gut
11. Eignung zum Feldflieger? ja, der geborne Jagdflieger

Einverstanden: ~~Schleißheim~~ L. Lechfeld, den 16. 6. 191

Egeler

Oblt. u. Führer der Flieger-Schule 4

Ibendinger

Leutn. u. Flugleiter

647/18.

Above: Uffz. Josef Bauer in the cockpit of his green and brown speckled Albatros D Va; his mechanics pose by the side of the aircraft. (G. VanWyngarden)

Bavarian Flieger-Ersatz-Abteilung 1 in Schleißheim. From January 18, 1918 onwards, he attended further training at Militär-Fliegerschule 4 in Lager Lechfeld. On July 2, 1918 he transferred to Armee-Flugpark 18 and was commanded to Jagdstaffelschule II in Nivelles from there. Less than three weeks later, on July 23, 1918, he was transferred to Jagdstaffel 76.[2]

Julius Geyer was born in Donauwörth on September 24, 1893. His father worked as the local district master builder. Initially, Julius worked as a merchant; from 1912 he began his 2-year military service with the Bavarian 4th Feldartillerie-Regiment. One year later, he was promoted to Gefreiter. Just prior to the conclusion of his military service, the war broke out. At the end of the first year of war he was promoted to Unteroffizier; at the end of 1917 he joined the Fliegertruppe. He was commanded to Bavarian Flieger-Ersatz-Abteilung 1 in Schleißheim and ended up at Militär-Fliegerschule 4 in Lechfeld in January 1918 for pilot training, along with Josef Bauer. He and Josef Bauer became friends; together they transferred to Jagdstaffelschule II via Armee-Flugpark 18 and ended up joining Jagdstaffel 76 together on July 23, 1918.[3]

Five days later, Leutnant Hans Hertrich arrived at the Staffel a second time. Everyone who still knew him from his first assignment with the Staffel was very surprised to see him again. They remembered Hertrich's nerves had been affected, and his fighter aviator skills had left much to be desired. Former Staffelführer, Lt. Walter Böning had clearly expressed his concerns in an assessment. Complaining was no use; the personnel department in Schleißheim had made their decision, and the Staffel had to come to grips with it.

By the end of the month, Uffz. Rupert Hainzl and Uffz. Edwin Hepperlin had arrived at the Staffel.

Rupert Hainzl was from Heining in Lower Bavaria (today part of the city of Passau) where he was born on March 25, 1898. Contrary to the other newbies at the Staffel, Hainzl had previous experience as a pilot on the front. He was a mechanical engineer by trade. He deployed to war with Prussian Flieger-Ersatz-Abteilung 1 in Altenburg in November of 1916. Initially a ground personnel member, he attended pilot training with Flieger-Ersatz-Abteilung 14 in Halle on the Saale River in June 1917. Upon successful completion he was commanded to Fliegerabteilung (A) 252 on January 21, 1918. Seven days later, he was transferred to Bavarian Fliegerabteilung (A) 286. On March 23, 1918 he was awarded the E.K. (Iron Cross) II and, on April 14, 1918 he received his Bavarian pilot badge. Following his promotion to Unteroffizier he transferred to Jagdstaffel 76 via Jagdstaffelschule II on July 29, 1918.[4]

In contrast to him, Unteroffizier Erwin Hepperlin, who had arrived at the Staffel on July 30, 1918, had no prior experience as a frontline pilot. He was born in Tuttlingen in Württemberg on January 3, 1896 and worked as a mechanic in Würzburg. Following the proclamation of mobilization, he joined the war as a volunteer like many others his age. He received basic infantry training with Bavarian 9th Infanterie-Regiment. In 1917 he decided to join the Fliegertruppe in order to learn how to fly. At the end of November 1917, he arrived at Bavarian Militär-Fliegerschule 4 Lager Lechfeld via FEA 1b in Schleißheim. He also attended Artillerie-Beobachter Schule in Grafenwöhr as part of his training. Upon completion of his flight training he arrived at Armee-Flugpark 6 on July 4, 1918. From there, he continued on to attend fighter pilot training at Jagdstaffelschule I in Valenciennes.[5]

Unfortunately, Erwin Hepperlin was also an example of the fact that not everyone who volunteered to become pilot was also suited for the task. In a statement written by Oblt. Amandus Rostock, the last Staffelführer, after the war, this situation was clearly described:

"To my regret I will be unable to issue the bravery card requested by Uffz. Hepperlin. After much consideration I have come to the conclusion that, if I am to honestly assess Uffz. Hepperlin's demeanor in the field, my conscience dictates the following evaluation. Hepperlin frequently displayed very little courage or fighting spirit which resulted in him being confronted about it almost on a daily basis, forcing me to threaten with his removal from the pilot roster in the end. All of the Jagdstaffel 76 pilots were in agreement about this assessment of Uffz. Hepperlin."[6]

As Staffelführer Lt. Ludwig Schmid's recovery did not progress as fast as had been hoped for, the decision had been made by higher positions to end Jagdstaffel 76's leaderless condition which had already lasted more than two weeks.

Above: Studio portrait of Uffz. Josef Bauer, this time in full flying gear and wearing his fur-lined coat.

According to an August 8, 1918 entry in the war diary of Royal Saxon Jagdstaffel 72, Leutnant Eberhard Grosse had been commanded on August

Above: Albatros D Va D.7544/17 served with Jagdstaffel 76, and is seen here in relatively new condition, before any unit markings were applied to the nose or tailplane. Note the rear-view mirror mounted in the cut-out of the upper wing. When this photo was taken, it appears that a Garuda propeller was fitted. This was one of the very last Albatros D Va to be manufactured, production of the type at Johannisthal ended with D Va D.7549/17. (V. Koos)

7, 1918 to assume the role of deputy leader of Jagdstaffel 76. He was born in Kammelwitz in the Breslau district of Upper Silesia on November 2, 1894. Before the war, Eberhard Grosse had joined the 5th Fußartillerie-Regiment in Posen as Fahnenjunker. When the war broke out, he deployed to the field with his unit. On January 31, 1915 he was promoted to Leutnant. He volunteered for service with the Fliegertruppe and began observer training with Flieger-Ersatz-Abteilung (FEA) 4 in Posen and FEA 1 in Döberitz on July 5, 1915 (the Jagdstaffel 76 ranking list notes August 7, 1915 as his start date). At the beginning of September 1915, following his graduation from observer training, he was initially assigned to Armee-Flugpark 6. On September 28, 1915 he was transferred to Feldflieger-Abteilung 24 as an observer. The following 12 months had been shaped by numerous temporary duty assignments to hot spots on the front. He did not only fly missions with his parent unit but also assisted with missions carried out by Feldflieger-Abteilung 62 and Artilleriefliger-Abteilungen 227 and 235. Early in September 1916, he began pilot training. Upon completion he transferred to Flieger-Abteilung (A) 264 on January 12, 1917. After he had volunteered for single-seater training in the fall of 1917, he attended fighter pilot training with the Jagdstaffelschule beginning on November 12, 1917, followed by his transfer to Armee-Flugpark 14 on the Italian Front, from where he joined Jagdstaffel 1. Upon his return to the Western Front, Leutnant Grosse joined Royal Saxon Jagdstaffel 72 on March 13, 1918.[7] This Staffel was under the command of Leutnant der Reserve Carl Menckhoff, who was soon to be decorated with the "Pour-le-Mérite". He was an extraordinarily experienced pilot who had achieved an aerial victory during his time at Jagdstaffel 72, and Jagdstaffel 76 was fortunate to have him as their new deputy commanding officer. His predominant task was to turn around the Staffel which was still equipped with outdated and worn-out Albatros D Va, and partially demoralized personnel.

Jagdstaffel 76 now had a Prussian Officer from the province of Silesia as Staffelführer who had been commanded to this temporary assignment from a Royal Saxon Jagdstaffel. Tasking an Officer who was not part of the Bavarian Army with deputy leadership of the unit was indicative of just how serious the lack of suitable potential Staffelführer was in the Bavarian Fliegertruppe at the time.

Uffz. Josef Bauer recounted the weeks he spent in the Champagne region:

"While other Staffeln, flying the Fokker, really

Above: This photo documents the less appealing condition of D.7544/17 after Uffz. Josef Bauer crashed it on 18 August 1918. By then, the unit marking had been applied to the horizontal tail, while the rear-view mirror has disappeared. This, coupled with the lack of personal markings, indicates that the aircraft served as a reserve machine by the time of the crash.

gave their opponents hell, we frequently failed to reach the necessary altitude in our old, worn-out Albatros. We felt like moving targets. Something was always broken on our aircraft. On every single flight we had to worry if the old crates were going to fall apart."[8]

The Jagdstaffel 76 pilots longingly eyed the new aircraft flown by the two other component units in the Jagdgruppe. Jagdstaffel 22 had taken delivery of Fokker D VII aircraft in July, while Jagdstaffel 63, which was stationed on the same field, received their new aircraft in early August 1918.[9]

Jagdstaffel 76 continued to fly their daily missions, but they did not achieve any aerial victories; fortunately, they also did not have any losses. In light of their outdated aircraft equipment and the largely inexperienced pilots in the pilot´s seats, this in itself was to be considered a success.

On August 11, 1918, Jagdstaffel 63 began to pack up in order to prepare for yet another move. They had been issued orders to transfer to the area of the German 6th Army.[10] One day later, Lt. Neu returned to Jagdstaffel 76 following recovery from his injury.

A few days later, new orders reached Jagdstaffel 76; they, too, were no longer needed in the Champagne region and were to relocate back to the Alsace instead.

On August 18, 1918 when the Staffel had already loaded the majority of their materials onto trucks, the aircraft were called for another mission on the front. A Kette of Albatros D Va departed under the Command of Lt. Grosse heading for the area near Chateau Thierry. They were still climbing over German territory when they were attacked by 12-14 enemy fighters who descended on them in a nosedive, separating the German Kette. Uffz. Josef Bauer's Albatros D Va D.7544/17 was attacked from behind by multiple enemy fighters, inundating his aircraft with a hail of bullets. Josef Bauer entered a spin in an attempt to lose altitude, but he continued to be pursued by the fighters; they continued to shoot at him and did not turn away until they found themselves in the focused fire of the German anti-aircraft gunfire. The Albatros was heavily damaged and hardly responded to movements of the control stick. Bauer feared that his aircraft would break up any moment. The engine, which had received multiple hits, finally quit, emitting a trail of smoke. Josef Bauer handled his aircraft like a raw egg as he descended toward a meadow in a glide. At low altitude, the plane abruptly plummeted down, impacting hard enough to shear off the landing gear. The Albatros slid across the ground, turned onto its nose and finally flipped on its back, coming to a rest.

Completely befuddled, Josef Bauer crawled out of the wreckage of his aircraft. He was relieved to

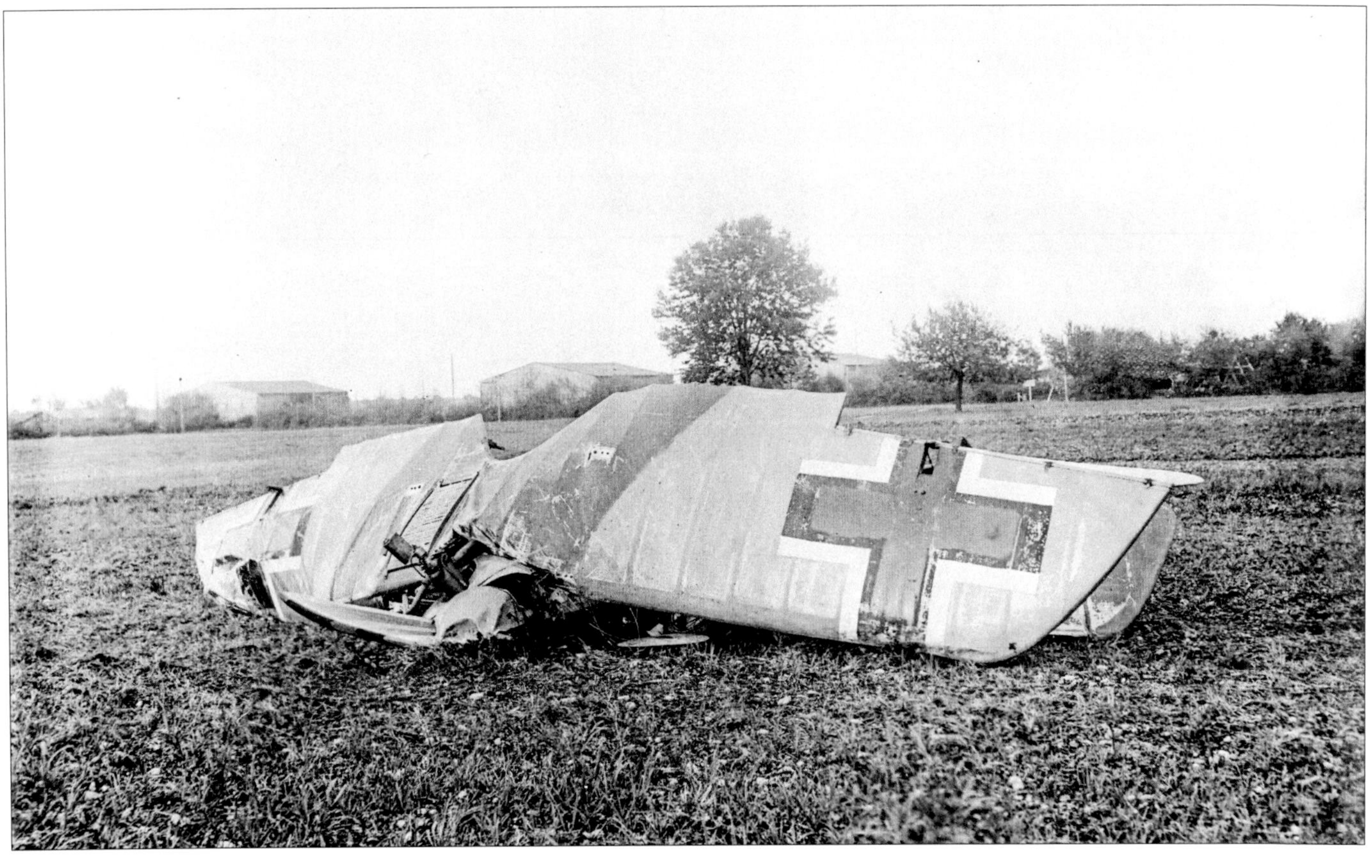

Above: The worn appearance of the painted camouflage on the wing upper surfaces is rarely seen in photos and may be of interest to the modeler. A close look at the conversion of the upper wing cross is also provided, and by the time of the crash a Heine or Niendorf propeller has been fitted.

discover that he was largely unharmed, with the exception of a few cuts and bruises. Meanwhile, German soldiers had appeared, and they were just as surprised to see him emerge from the smashed up aircraft without any visible injury. A soldier offered him a cigarette. Bauer gladly accepted, leaned against the fuselage of his Albatros and enjoyed a smoke. Since the aircraft was a complete write-off, the imminent risk of fire could cheerfully be ignored in this case.[11] Later, Clemens Vollert arrived in the Staffel car. He had brought his camera, capturing images of the destroyed Albatros D Va. Vollert told Josef Bauer that all other pilots had safely made it back, even though most of them had received multiple hits to their aircraft. This was the last mission flown by Jagdstaffel 76 in the area of the German 1st Army.

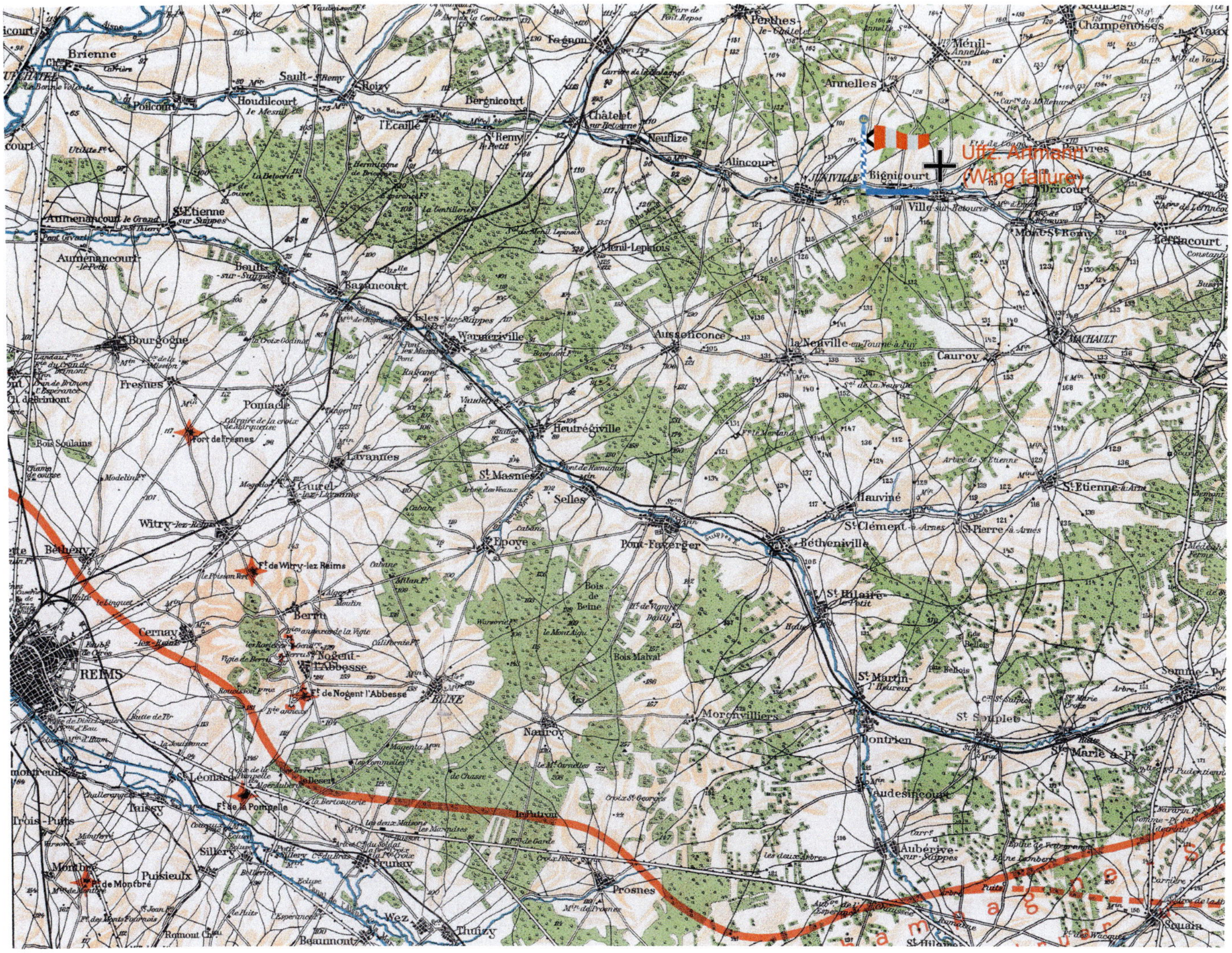

Bignicourt Airfield
Location & Events

#		Date	Pilot	Location	Footnote
1.	+	12.07.18	Uffz. Artmann	Bignicourt	KIC (Wing failure)
2.	+	16.07.18	Lt. Schmid	?	WIA
3.	+	16.07.18	Lt. Neu	?	WIA

6 Back at the Old Base in Upper Alsace

German Armee-Abteilung B, August 20, 1918 until the End of War

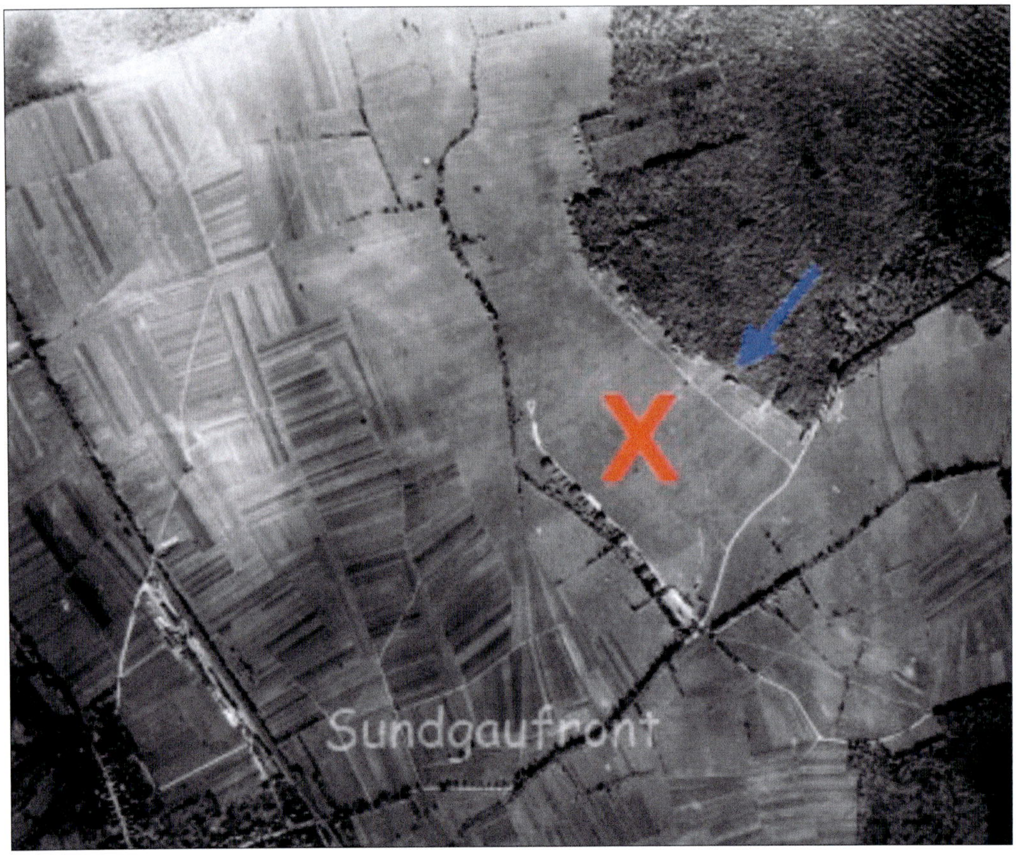

Left: A small aerial photo of Sierenz airfield, marked with an "X". At right, by the edge of the forest, the hangars of the Jagdstaffel can be seen. On the left the avenue that can also be seen in photo 3 can be made out.

After a two-day railway trip, the Staffel returned to Sierenz airfield on August 20, 1918. The airfield was situated approximately 12 km south-southwest of Mulhouse, about halfway between their former base at Habsheim airfield and the border to Switzerland. Fate had brought them right back to the area where the unit began frontline operations approximately nine and a half months earlier.

After relocating to their new airfield, the Staffel was finally given the opportunity to replace their completely worn-out Albatros D Va aircraft for a new type. Much to the pilots' disappointment however, these aircraft were not replaced by the formidable Fokker D VII, the type they had longed for. Instead, numerous examples of the Pfalz D XII were delivered, which was a new but unpopular type of fighter.[1]

Nevertheless, the pilots were relieved to have newer equipment at their disposal, and that they would no longer have to concern themselves with the fear of their aircraft breaking up in mid-air. Even though they had not received Fokker D VII, the new "birds" were a significant improvement compared to their old Albatros. The Staffel conducted routine flights to the front in their Pfalz D XII, none of which yielded any noteworthy results, however.

On September 4, 1918 after more than a month, the "2nd Hertrich era" had come to an end for the Staffel, as Leutnant Hertrich returned to Schleißheim. In a follow-up to Lt. Hertrichs second departure from the Staffel, the deputy Staffelführer Leutnant Erhard Grosse composed the following "case" analysis on September 9, 1918:

"Leutnant d.R. Hans Hertrich had first transferred to the Staffel as a pilot on February 18, 1918. His flying skills did not meet the requirements expected of a fighter pilot. He himself recognized this fact, requesting his replacement; on April 6, 1918 he was transferred to the Park and back home from there. On July 27, 1918 he was transferred

Above: Jagdstaffel 76 at Sierenz airfield, with the newly equipped with Pfalz D XII aircraft. Tarp covers protect the props and tires against the hot summer sun, and the engine side panels, and nose chin panels have been removed as well. The setting indicates that these photos may have been taken around August 21, 1918, during the last hot days of the summer. (R. Absmeier)

Above: Rear three-quarter view of the same line-up of Jagdstaffel 76 Pfalz D XII aircraft at Sierenz. The first aircraft on the left is the Pfalz D XII of Uffz. Bauer. (G. VanWyngarden)

back to the Staffel as a pilot - surely, this transfer was an erroneous act. Upon a request issued by Leutnant d.R. Hertrich himself, he was made available to Iluft on September 1, 1918. As a result of his short frontline service, he had not gained a lot of experience. It is unlikely he would meet the mental and physical demands of a front pilot as his nerves seem spent; he would serve his duty well in the position of instructor, technical officer, or flight instructor. Quiet and modest officer.

Grosse

Leutnant and deputy Staffelführer"[2]

On September 17, 1918, Jagdstaffel 76 received orders that they were to be combined with Jagdstaffeln 69, 71 and 75 to form a Jagdgruppe under the command of Oblt. Thomas, Staffelführer

Above: Uffz. Josef Bauer in the cockpit of his new Pfalz D XII 1462/18, with a signal pistol and flares mounted externally below the cockpit. His ground crew gather in front of the aircraft.

of Jagdstaffel 69. The following day, orders were issued to relocate to Habsheim. The aircraft and all equipment were loaded onto trucks in order to be moved the short distance to the new airfield.

With the formation of the Jagdgruppe, and much to their pleasant surprise, the long-anticipated Fokker D VII finally arrived in exchange for the unpopular Pfalz D XII. Four of the fairly new Pfalz D XII fighters were retained by the Staffel as reserve aircraft.[3]

On that same day, Lt. Obermeier and Gefreiter Reinhard and Brunner received their Bavarian pilot badges.

On September 20, 1918 Julius Geyer wrote a letter home:

"My God, why did we not receive these aircraft sooner? The damage we could have done with them on the front (the 2nd and 1st Armies, author's note)! We can hardly wait to fly as often as possible!"[4]

On October 7, 1918 Oberleutnant Amandus Rostock arrived as the new Staffelführer; he was transferred from Royal Bavarian Jagdstaffel 77. Staffelführer Lt. Ludwig Schmid, who was still absent due to continued complications related to his injuries, had been officially transferred to Bavarian Flieger-Ersatz-Abteilung 1 in Schleißheim the day prior. This concluded Lt. Eberhard Grosse's time with the Staffel. He was sent off with a festive social event organized by "his Bavarians" before he returned to Jagdstaffel 72.

The Jagdstaffel 76 ranking list includes the following remarks: *"Lt. Grosse acquired a great many merits during his time as deputy Staffelführer."*[5]

Amandus Rostock was born in Freienorla (on the Saale River, south of the city of Jena in Thüringen) on March 6, 1895. He was the son of a landowner. After graduating from a humanistic high school, he studied math and law at Würzburg University for one semester each. Thereafter, he fulfilled his military service duties as a war volunteer with the Bavarian 9th Infanterie Regiment in Würzburg. He

Above: The Uffz. Julius Geyer in the cockpit of his Pfalz D XII 1516/18, marked with his personal insignia, the ribbon of the Iron Cross 2nd Class, barely visible at the left edge of the image. Different to Bauer´s Pfalz, the flares are mounted to the fuselage with the help of a fabric belt.

was discharged into civilian life holding the rank of Unteroffizier and Offizier-Aspirant der Reserve. He began vocational training as a trainee with Dresdner Bank. The outbreak of the war brought a sudden end to his training. Following mobilization, Amandus Rostock was drafted for duty with Bavarian Reserve-Infanterie Regiment Nr. 4 on August 4, 1914. He moved into the field as a member of the Regiment and was stationed at the Vosges front. After taking part in fighting near Sennheim, he was wounded in his hip during a further border on August 27, 1914. Following his recovery, he returned to his Regiment, which was now fighting on the Flanders front.

On March 18, 1915 he was promoted to Leutnant der Reserve. As he was seeking a new challenge, he volunteered to join the Fliegertruppe, in order to become an observer. His request was granted, and he began observer training with Bavarian Flieger-Ersatz-Abteilung in Schleißheim on May 27, 1916. After he completed the first part of his training, he was transferred to Armee-Flugpark 6, where he completed the remainder of observer training which lasted several weeks. After graduating, he transferred to Bavarian Feldflieger-Abteilung 2 on August 21, 1916. On March 21, 1917 following just over 7 months of duty as an observer, he returned to Schleißheim where he served as a bomb instructor at the observers training facility.

In late October 1917, he began pilot training with Militär-Fliegerschule 1. After completing 51 training flights, he took off on his first solo flight as early as November 23, 1917 and he passed all subsequent exams with distinction. His training was completed by a temporary duty assignment with Jagdstaffelschule 1 where he attended an additional single-seater course. On May 4, 1918, now promoted to Oberleutnant, he arrived at Royal Bavarian Jagdstaffel 77. On May 15, 1918, Oblt. Rostock stepped in as deputy Staffelführer to take over the duties of Oberleutnant Walter Ewers who had fallen on the same day. Oblt. Rostock held the position for one week until Lt. Rudolf Stark took over as deputy Staffelführer on May 23, 1918.[6]

Lt. Max Gossner, who would later be his Staffelführer at Jagdstaffel 77, issued the following assessment about him:

"…. Oberleutnant Rostock has been a pilot with Jagdstaffel 77 since May 4, 1918. During his time with the Staffel, he achieved two aerial victories. His off-duty demeanor is flawless. Oblt. Rostock is an ambitious, skilled Officer with a considerable general and military education.

Above: Members of the ground crew pose in front of a Pfalz D XII at Sierenz, in a setting that is remarkably similar to the photos taken early in the year front of the Albatros fighters at Habsheim, seen at the end of Chapter 3 of this book. (G. VanWyngarden)

Below: Jagdstaffel 76 mechanics in front of a Pfalz D XII, shortly after the return to Habsheim. Pictured in the centre is the Feldwebel who had been recruited by Walter Böning while the Staffel was formed at Habsheim. He was instrumental in training the ground crew into the group of efficient servicemen that they were. He continued to lead the mechanics until the end of the war.

Above: The Fokker-built D VII that was the personal aircraft of the new Staffelführer Lt. Eberhard Grosse. The tail of an O.A.W.-built D VII of the unit can barely be seen in the background. (G. VanWyngarden)

His flying skills are good.
I consider Oblt. Rostock suited to serve as Staffelführer.

Signed. Max Gossner, Leutnant and Staffelführer."[7]

Otto Fuchs who was a member of Jagdstaffel 77 along with Amandus Rostock shared the following memory of his comrade:

"He was a career officer but no "Kommisskopf" (Army head/mind), but was fairly ambitious. With his comrades he was laid back and always up for a good joke. He was well-liked among his comrades, as well as the ground personnel. He was a good pilot and had an eye for aerial battles because he was an observer before he became a pilot. This attribute made him a reliable comrade and a good Kettenführer. He displayed ambition and courage without being boisterous, weighing his options prior to each attack. I could envision him as a good Staffelführer. His nickname with Jagdstaffel 77 was "Schneckerl" due to his golden blond curls (in Bavaria and Austria this was a common expression used for people with curly hair, author's note)."[8]

Whenever circumstances allowed, the Jagdstaffel 76 pilots took their prized new fighters aloft. They encountered French aircraft on several occasions but initially, none of them yielded any successes, but no losses were suffered either. While the motivation of the pilots was high after converting to the Fokker D VII, they were unable to achieve much in the remaining weeks of war as there were now issues with aviation fuel supply. According to a report by Jagdstaffel 71 pilot Hans-Joachim von Hippel, only a limited fuel supply was available for aircraft beginning in October. As a result, each Jagdstaffel only flew on certain weekdays that were assigned to them by the Commander of the Jagdgruppe.[9]

On October 30, 1918, they finally achieved a success. A Staffel Kette, composed of three Fokker D VIIs, attacked a French Salmson two-seater (referred to as D.H. 9) from Escadrille Sal. 41 near

Above: Oblt. Amandus Rostock takes a moment to relax outside while serving as a pilot of Jagdstaffel 77 at Foucaucourt-en-Santerre airfield. The sign in the back shows the Bavarian coat of arms. His nickname was "Schneckerl" (curly), obviously inspired by his curly blond hair.

Above: Portrait of Oblt. Amandus Rostock, drawn by former Jagdstaffel 77 member Eduard Illig, who was a professional portrait painter, on October 14, 1918. Since the portrait was dated a week after Rostock was posted to take over command of Jagdstaffel 76, it was probably intended as a farewell present by his former Staffel comrade.

Altkirch. Lt. Karl Dumle succeeded in forcing the French aircraft to land west of Carspach in French lines, scoring his first and only aerial victory in the process.

The report in the French daily Communiqué read:

"Attaqués par 3 avions Fokker, le sergent pilote Lotrat est légérement blessé à la tète par eclats de balles et le mitrailleur soldat Monjalat est grièvement blessé à la cuisse par une balle."

"Attacked by three Fokker aircraft, pilot Sergeant Lotrat sustained light injuries to the head caused by bullet splinters; rear gunner Soldat Monjalat was gravely injured by a shot to his upper thigh."[10] This also turned out to be the very last combat success achieved by Jagdstaffel 76 during the course of the war. It should be noted that Offiziere Maximilian Veith and Karl Dumler received the Bavarian pilot badge, and Gefreite Geyer, Reinhard, and Brunner were promoted to Unteroffizier.

As no events worth mentioning occurred on the Alsace front during the last two weeks of the conflict, one would think that all important aspects concerning Jagdstaffel 76 have been covered, but that is not the case: the history of Jagdstaffel 76 still holds one last small chapter which casts a light upon the chaotic conditions that prevailed during the last days of war.

In the meantime, Walter Böning had finally made a full recovery and was getting rather bored at Militär-Erholungsheim (military rest home) Bad

Above: The new Staffelführer Oblt. Rostock, 3rd. from left, poses for a group photo together with his mechanics in front of his newly delivered Fokker D VII (O.A.W.) 6372/18. The tail of the Fokker D VII (O.A.W.) flown by Uffz. Bauer can just be seen below the fuselage of Rostock´s aircraft.

Reichenhall. From There, he had sent out repeated requests to be put back in charge of Jagdstaffel 76.

"I was very much aware that we would not end up winning the war, but I wanted to return to my Staffel and make sure that the boys would safely make it through the last weeks. Not being able to do anything to help them worried me sick."[11]

On November 1, 1918 he was instructed to return to Habsheim where he would be reunited with the Staffel. He wrote about the events in a letter to his family, dated November 11:

"I was at Reichenhall on November 1; I drove to Mulhouse via Munich etc., and arrived in Mulhouse on the 3rd (of November) around noon. My transfer should have been made known out there immediately upon my arrival, as an Oberleutnant, who had arrived as new commander eight days ago, would hardly have a reason to leave otherwise."[12]

When Walter Böning arrived back at Jagdstaffel 76 on November 3, 1918 he was warmly greeted by Clemens Vollert, Josef Diem, and the rest of the non-commissioned officers whom he had recruited in Schleißheim back in the days of the formation of the Staffel. When he entered his office, he was greeted by a perplexed Oblt. Amandus Rostock, who was totally unaware of Lt. Walter Böning's (re-) appointment as Staffelführer. Multiple inquiries with Armee-Oberkommando and Kommandeur der Flieger returned that no such orders from Munich were known. Without having received such an order, Oblt. Rostock could not and would not hand over leadership of the unit to Walter Böning, who fully understood the situation his colleague found himself in.

Walter Böning remained at Habsheim for another three days. He closely examined the Staffel's new Fokker D VII and could not resist the temptation of undertaking a few flights on a Fokker across the Vosges with his old comrade Josef Diem. A thick cloud cover, poor visibility and rainy intervals

Above & Below: Fokker D VII´s at long last! Lt. Walter Curtius enjoying the view from the roof from one of the hangars at Habsheim airfield in October 1918. A full dozen of new Jagdstaffel 76 Fokker D VIIs are neatly lined up below him, the props having been carefully aligned horizontally to celebrate the occasion. Four remaining Pfalz D XII have been unmercifully shoved off to the distant end of the line-up. This photograph documents another German Jagdstaffel being fully equipped with the nominal number of aircraft (14+2 spares) shortly before the end of the war.

Above: Oblt. Rostock's Fokker D VII (O.A.W.) 6372/18, now sporting its full paintjob. The broad bar painted between the crosses on the upper wing identifies the aircraft as that of the Staffelführer.

resulted in low flying activities, both on the German and French side of the lines. As a consequence, none of these frontline flights resulted in contact with enemy aircraft.[13] As he would later recall, the Fokker D VII was the best aircraft he had ever flown.

He was happy to be back among pilots. Between the few missions flown he talked to the pilots and exchanged memories with Clemens Vollert and Josef Diem. His relationship with Amandus Rostock was good as he did not interfere with his way of leading the Staffel. In turn, Rostock would grant him free hand when he wanted to fly to the front. When no orders for the re-assignment of the Staffel under his command had been received by November 6, 1918, in spite of multiple inquiries with Armee-Ober-Kommando, he got on his way to return to Munich, feeling considerably annoyed. He was certain that there would be no path to friendship between him and the Bureaucracy of the Bavarian Military:

*"As a result of some inexplicable events the order did not arrive out there (*in Habsheim, author's note*) until the 7th. I had already returned to Munich on the 6th in order to find out what was in store for me. When I arrived in Munich, I was told to immediately pack my belongings and drive to the Staffel. I had to return to Reichenhall in order to get my luggage (in spite of his instructions, his luggage had not been sent from Reichenhall to Habsheim, author's note). I returned to Munich on the 8th (in order to continue to Mulhouse from there, author's note). At the Freilassing train station (a village east of Munich), people were selling horses, wagons, cannons, guns, etc. which had been left behind by Austrian troops..."*[14]

Once the train reached Mühldorf, it was stopped by revolutionary soldiers:

"Suddenly the door was swung open, two soldiers stood in front of me. One of them pulls the cap from my head and rips down my cockade. I punch him (a boy of maybe 18 or 19 years of age) in the mouth hard. Shots fired through the window. His comrade had missed even though he only stood three steps from me. Then, they sobered up and went their way."[15]

As there were no more trains leaving Mühldorf for the time being, he had to abandon his plans to travel

Above: A smiling Uffz. Julius Geyer seated in the cockpit of his new Fokker D VII (O.A.W.) 6381/18. The fuselage has again been marked with his personal insignia, the Iron Cross 2nd Class fuselage band. Geyer has also expanded the unit marking on the tailplane somewhat by painting the rear fuselage below the horizontal tailplane blue. The military number of this aircraft can be deciphered on the forward landing gear strut.

to Mulhouse in order to take back the leadership of his Staffel. However, Walter Böning finally managed to make his way back to Schleißheim by November 11, 1918 via numerous detours. Upon his arrival, he was surprised to learn that he had been elected "Führer des Soldatenrates" (Lead of the revolutionary Soldier's Council) as a result of an initiative led by multiple Unteroffiziere. The next day, he received orders to stay in Schleißheim as Demobilisierungsoffizier (Demobilization Officer).

At the same time Walter Böning had arrived in Schleißheim, the war was also over for his former Staffel at Habsheim airfield – as it was for everybody else who was involved in the conflict. Oblt. Amandus Rostock and Lt. Clemens Vollert led the Staffel personnel back to Schleißheim unanimously, where the Staffel was demobilized. The Staffel aircraft were left behind at Habsheim and were taken over by Entente forces that took over the premises. Some of the Fokker D VIIs formerly operated by the Staffel are documented to have come into use with the US Expeditionary Forces at Trier at a later date. Several were also photographed in storage at Romorantin, where captured aircraft were gathered in order to select those suitable for transport to the USA.

On a last note, Walter Böning had been working as Demobilisierungsoffizier in Schleißheim until the day of his release from the military on January 10, 1919. In his function he encountered his former Staffel again. He met up with Clemens Vollert, Josef Diem, and the Staffel Unteroffiziere who he had once recruited in Schleißheim. They spent a few memorable evenings at a local tavern exchanging memories before each of them went their separate ways.

Walter Böning later explained that he could not understand why Oblt. Rostock did not fly the Jagdstaffel 76 aircraft back to Schleißheim. He stated the following on the situation: *"I think Rostock was ordered to leave the aircraft behind at Habsheim in order for them to be taken over by the Allies. No matter what the orders would have been, I*

Above: Blow-up of the Fokker D VII (O.A.W.) flown by Uffz. Bauer. A flare pistol has again been mounted below the cockpit opening, just as on his previous Pfalz D XII.

Above: Officer pilots of Jagdstaffel 76 gather for a photo at their Habsheim casino in October 1918. From left: Lt. Martin, Oblt. Rostock, Lt. Neu, Lt. Vollert, Lt. Dumler, Lt. Obermeyer, Lt. Curtius.

Above: Close-up of the Fokker D VII (Alb.) flown by Lt. Karl Dumler, marked with a white wave band. The white square at the right tip of the marking is the factory-applied rigging diagram.

Below: Lt. Karl Dumler seated in the cockpit of his new Albatros-built D VII at Habsheim. When the photo was taken, the upper engine panels were not in place. Note the thick padding mounted to the rear ends of the machine guns.

Above: Fokker D VII (Alb.) 5341/18 flown by Lt. Maximilian Veith can be seen in the centre of this enlargement, and the Fokker D VII (O.A.W.) flown by Lt. Neu can be seen to the left of it. A close inspection of the picture proves that the planes had the letters "V" (Veith) and "N" (Neu) painted on the centre sections of their upper wings.

would have most certainly tried to fly them back to Munich."[16]

During its nearly one-year existence, Royal Bavarian Jagdstaffel 76 achieved at least 20 aerial victories (of which 19 had been officially confirmed by Kogenluft). Two more were only acknowledged as "jenseits der Linien zur Landung gezwungen" (forced to land beyond enemy lines). In the same period of time, three pilots had been killed in aerial combat, three pilots crashed fatally, and three were taken prisoners of war. An additional four pilots were wounded in aerial combat.

Lt. Walter Böning was the Staffel's most successful fighter pilot; he had achieved 11 aerial victories with the unit (out of 17 in total), Uffz. Siegfried Walther had two aerial victories, and Uffz. Diem one aerial victory, as well as one aircraft forced to land

Considering that the Alsace region was generally considered to be a "quiet" sector, the 12 confirmed aerial victories scored there were a considerable achievement for Jagdstaffel 76. At the hot spots of the front (2nd and 1st Armies) however, big achievements were hardly possible due to the entirely inadequate aircraft equipment and, at times, highly inexperienced pilots that had to fight an enemy who was considerably superior in numbers. Under these considerations the 8 aerial victories achieved there (in addition, one enemy aircraft forced to land) were to be weighed a lot more heavily and deserve special recognition.

Above: Pilots of the Staffel enjoy a cigarette or two behind the tail of Lt. Neu's Fokker D VII (O.A.W.). The letter "N" has been painted on the upper wing center section, and a rear-view mirror has been fitted to the wing cut-out. Note that the horizontal tail surfaces have not yet received the usual blue paint yet, only the white stripes have been applied when the photo was taken. Seen from the left are: unknown, Lt. Vollert, Lt. Lt. Obermeyer, Lt. Martin, Lt. Dumler and Lt. Curtius.

Below: The war is over! Jagdstaffel 76 members gather at Habsheim for a farewell photo before their return transport to Schleißheim (T. Weber)

Above: Fokker D VII (Alb.) 5341/18 formerly flown by Lt. Maximilian Veith is seen here in American hands at Trier after the armistice. A replacement rudder taken from a Fokker-built D VII has been used on this machine, and this component was already fitted to the plane when the line-up picture at Habsheim was taken. The rear-fuselage horizontal stripe was also applied to the rear of the fuselage decking. The rib tapes on the upper wing were of a light-colored material, shown as light blue in our profile. On the lower wing, the rib tapes were of four- or five-color fabric. (J. Leckscheid)

Above: Left-side view of Veith´s Fokker D VII (Alb.) 5341/18 after the war. Note the low position of the weights table and the fuselage datum line. The rear fuselage of Lt. Franz Büchner´s (Staffelführer of Jasta 13) last Fokker D VII (O.A.W.), complete with his personal lion´s head marking, can be seen at left.

Above: Fokker D VII (O.A.W.) 6481/18 of Jagdstaffel 76 at Romorantin, after the war. The military number of the plane can barely be made out on the rudder and fin. (G. VanWyngarden)

Above: This post-war hangar view of the Fokker D VII (O.A.W.) 6481/18, of Lt. Curtius illustrates the inboard position of the lower wing crosses on this late-production O.A.W.-built D VII. The white stripe of the unit tail marking has also been applied to the lower surfaces of the horizontal tailplane.

Above: Frontal view of Fokker D VII (O.A.W.) 6481/18 in the hangar at Romorantin. The landing gear and fuselage are covered with mud, indicating that the plane was flown recently. (G. VanWyngarden)

Above: This close-up of the previous photo shows the "sunburst" marking painted onto the lower nose panel. The horizontal water pump of the Mercedes engine shows nicely, the propeller is made by "Axial" but lacks the famous company decal, as did most of their late-war products. (G. VanWyngarden)

Above: The welcome of the Koller family for Karl Koller when he returned from British captivity on October 22, 1919, after nearly 1½ years as prisoner of war. (Koller)

Above: Fokker D VII (O.A.W.) 6423/18 of Jagdstaffel 76 after the armistice at Trier. Note the small windscreen fitted by the new owners. This is another case of the horizontal tailplane not being painted with blue paint, the polygons of the five-color fabric covering this component are still visible in this view. (G. VanWyngarden)

Above: An American pilot in the cockpit of Fokker D VII (O.A.W.) 6423/18. (G. VanWyngarden)

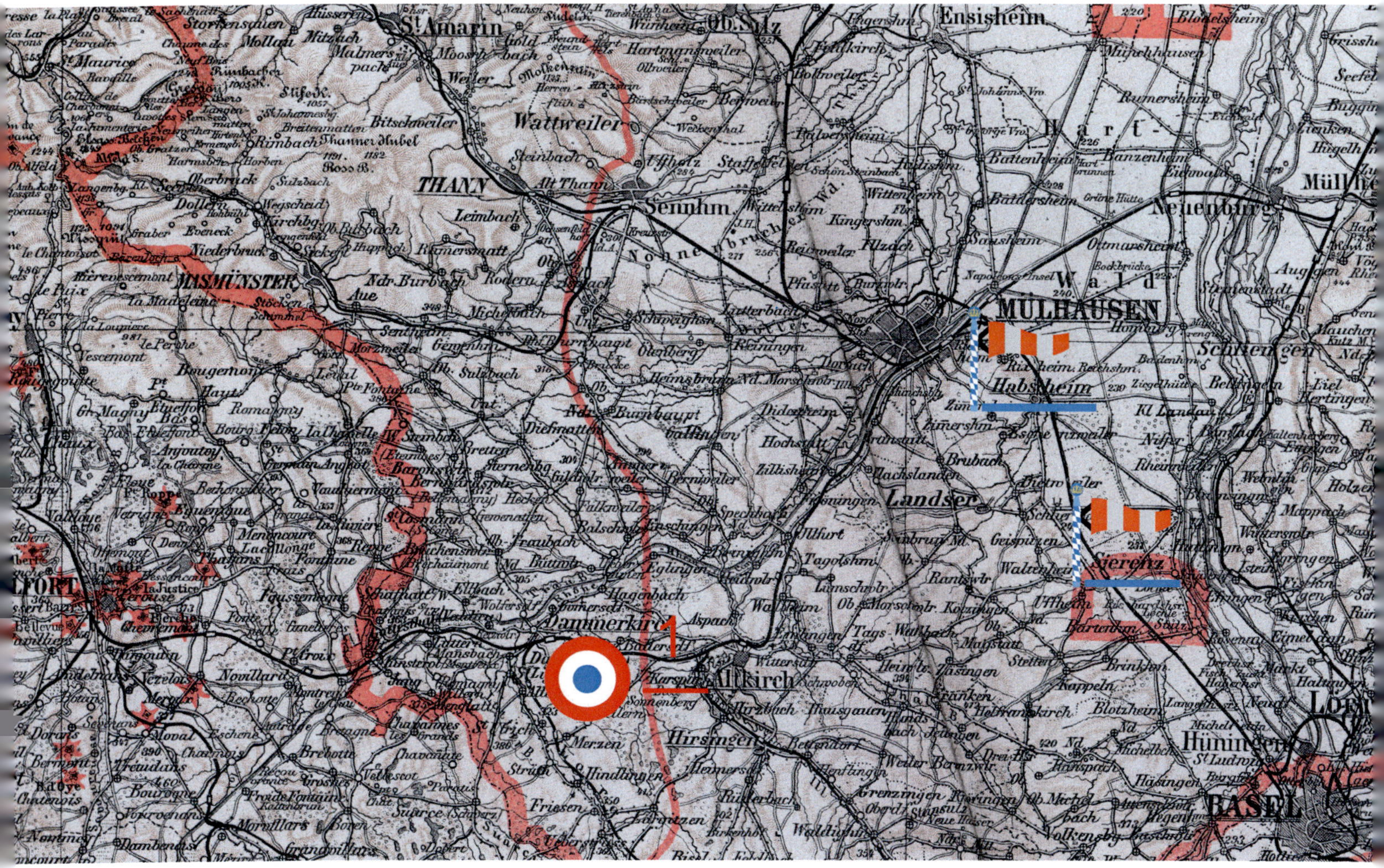

Habsheim Airfield
Location & Event

#		Date	Pilot	Location	Footnote
1.	◎	30.10.18	Lt. Dumler	West of Carspach	Salmson 2A2

7. Aircraft Markings and Paint Schemes

Above: The Bavarian white over blue flag, used as Staffel-marking of Jagdstaffel 76.

Right: The Uniform of the Bavarian 4th Chevauleger Regiment, in this example the uniform of a trumpeter.

Whenever I had the opportunity to interview the old pilots over a timespan of several days, I always refrained from asking questions concerning the markings and paint schemes of their aircraft right at the start at the conversation. Instead, I only touched this subject towards the end of the interviews. The reason was that at this stage my interlocutor´s memory had arrived back in the time when he was a fighter-pilot in World War I. In some cases, they literally re-lived the adventures of their wartime flying days, and when looking again at old photos at this stage, surprisingly detailed information on the colors of the aircraft had re-emerged.

My conversations with Walter Böning extended over a timespan of four and a half days, and accordingly left plenty of room to talk about the paint schemes and markings of the aircraft. For this reason, all colour profiles of the Albatros D III and D Va presented in this book were created on the basis of his information, unless another source of information is referred to in the footnotes.

Further profiles were created on the basis of documents and notes found in the possession of former pilots, or from information gathered from official documents. Others were created on the basis of documents in the bequests of other former pilots or official documents. It is in the nature of historical research that, in some cases, the colors for the artwork of aircraft had to be extrapolated. This is especially true for aircraft used by the Staffel towards the end of the war. In these cases, the source of information is noted in the text.

1. Staffel marking of Albatros D III (O.A.W.) and D Va aircraft. (Top-view tail)

The Albatros D III and D Va noses were painted white and blue for identification; the propeller spinner was white. Top and bottom surfaces of the horizontal tail planes were painted in Bavarian blue with a white lengthwise stripe centred on each side. Dr. Böning noted that as a Staffelführer of a Bavarian Staffel, it was of course necessary to display the

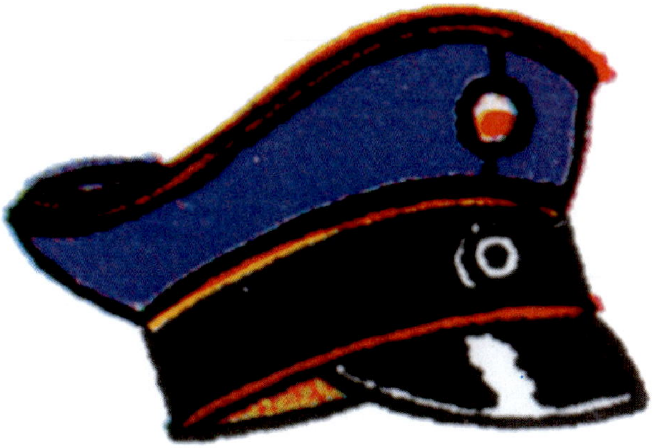

Above: Military cap of the Bavarian Telegraph Battalion.

Above: The flag of the city of Munich.

Below: The "Rauten" Version of the Bavarian Flag.

colours of the Bavarian Flag. The horizontal tail surfaces were an ideal position for the placement of a Staffel marking. On patrols, the aircraft flew in a V-shaped formation and staggered in altitude, thus the tail was always well visible.

As an additional note, the wings of all Albatros D III (O.A.W.) shown in the profiles were finished in five-color dyed fabric. The Albatros D Va could be either finished in this material or the usual two-color painted camouflage upper surfaces and light blue below – see the profiles. Profile 1 on p.204

Above: The coat of arms of the Hochstift Würzburg (central administration of the archdiocese of Würzburg).

2. & 2a.Lt. d. Res. Walter Böning, Albatros D III (O.A.W.)

In addition to the Staffel identification markings, his aircraft (and all of his other aircraft) bore his personal insignia, a blue-white-blue band painted around the fuselage, between the cockpit and the fuselage cross. The horizontal tail surfaces, as well as the rudder, were painted "baby blue", to use his own wording. After the Staffel had been equipped with the Albatros D Va, he only flew his Albatros D III (O.A.W.) as a reserve aircraft until it was finally returned to the Armee-Flugpark.

3. Lt. d. Res. Franz Xaver Danhuber, Albatros D III (O.A.W.)

In addition to the Staffel identification markings, Xaver Danhuber's Albatros D III (O.A.W.) displayed an oblique black and white band around the fuselage, below the cockpit. The tail section was painted black headed by a thin white stripe, a reference to his previous unit, Jagdstaffel 26. The aircraft operated by his former unit had been marked with alternating black and white fuselage bands. Fin and rudder were painted blue. The aircraft was damaged during take-off on November 23, 1917, by Vzfw. Walther.

4. Vzfw. Paul Wagner, Albatros D III (O.A.W.)

Above: Military cap of the Infantry.

In addition to the unit markings, the fuselage of Vzfw. Paul Wagner's Albatros, including fin and rudder, was painted green. This color was chosen by him as a reference to his former unit, the Bavarian 4th Chevauleger-Regiment (light cavalry), which traditionally wore green uniforms.

5. Vzfw. Siegfried Walther, Albatros D III (O.A.W.)

The fuselage of Vzfw. Siegfried Walther's aircraft tail was largely painted red, except for the metal engine covers and the fuselage bottom. In early December 1917, he crash-landed this aircraft at Habsheim airfield. The aircraft was a complete write-off and sent back to the Armee-Flugpark for salvaging whatever components were suitable for future use. The reason why he adapted red as his personal color is not documented, but since he was a member of the Infantry before joining the Fliegertruppe this very likely could be a reference to his parent unit, which generally wore red cap-bands. The photos taken after the crash document that no unit markings had yet been applied to the nose or horizontal tail surfaces, indicating that the crash occurred soon after the aircraft reached the unit.[1]

6. Oblt. Erwin Wenig, Albatros D III (O.A.W)

Oblt. Erwin Wenig's Albatros D III (O.A.W.) was marked with a black fuselage, fin and rudder. As Erwin Wenig later verifiably used the color black as his personal insignia while serving as Staffelführer for Jagdstaffel 80, it is likely, though not certain, that this Albatros D III was his personal aircraft. Even back in the days of his service as a Fokker-flyer he had adopted a black fuselage band as his personal insignia. The reason for this is most likely that he chose it as a reference to his parent unit, the 1st Bavarian Telegraph Bataillon, which wore black cap bands. The aircraft's horizontal tail surfaces, as well as the nose, were painted in Staffel colors. The

4. Infanterie-Rgt.
Infanterist

Above: Uniform of a soldier of the Bavarian 4th Infantry Regiment.

Above: Flag of the Prussian province of Silesia.

identity of the pilot who took over this aircraft after Wenig was transferred has not been recorded.

7. Vzfw. Karolus Bärenfänger, Albatros D III (O.A.W.)

Karolus Bärenfänger´s first personal aircraft as a member of Jasta 76 was this Albatros D III (O.A.W.). The fuselage of this machine was painted with longitudinal white and blue stripes, replicating the colors displayed on the Bavarian flag. Similar white and blue stripes were applied to all the aircraft he flew during his service with Jagdstaffel 76.

8. & 8a. (Top-view) Lt. d. Res. Walter Böning, Albatros D Va

In addition to the Staffel identification markings, his Albatros D Va was identified by a "baby-blue" fuselage and elevator surfaces. The rudder was painted white. His personal insignia on this aircraft was his usual blue-white-blue ribbon. Walter Böning recounted that initially he had the idea that the "baby-blue" fuselage should have been applied to all aircraft of his Staffel. But the idea was dropped because pilots preferred choosing their own colours, making it easier to recognize the individual aircraft in aerial combat.

9. & 9a. Lt. Hans Böhning, Albatros D Va. D.5727/17

In addition to the standard Staffel identification markings, Lt. Hans Böhning's Albatros had white and blue (Rauten) lozenges painted onto the rear fuselage behind the cockpit, headed by a narrow black band. This paint scheme presented the most popular version of the Bavarian flag, the black stripe in front of the "Bavarian Rauten" was a reference to his parent unit, the 10th Royal Bavarian Field-Artillery-Regiment, which had black cap bands. The Albatros D V he flew before at Jagdstaffel 36 was marked with an almost identical paint scheme. The oft-repeated claim that this aircraft belonged to Walter Böning is incorrect; it can be traced back to a mix-up of names by an aviation historian in the 1960s.[2]

10. Lt. d. Res. Wilhelm Kolb, Albatros D Va D.5294/17

His Albatros D Va had a "baby-blue" fuselage, like the one flown by his Staffelführer and friend Walter Böning. The personal insignia applied to this aircraft was a two-color band wrapped around the fuselage, representing the city colours of his hometown of Munich, black and yellow. Nose and horizontal tail surfaces of this Albatros were painted in Staffel identification colors.

11. & 11a. (Top-view) Vzfw. Karolus Bärenfänger & Lt. Moritz Weig Albatros D Va

The fuselage of his Albatros was painted in his usual personal markings, blue and white stripes painted horizontally along the fuselage. The hubcaps of the aircraft were also painted in blue and white. Nose

Above: Military cap of the artillery.

Above: Uniform of the Bavarian 3rd Field Artillery Regiment.

and horizontal tail were painted in Staffel colors. After he obtained a new Albatros D Va, his aircraft remained with Jagdstaffel 76 and continued to serve as a reserve aircraft. While serving in this function it was flown by various pilots. Lt. Moritz Weig flew this aircraft when he when he was obliged to make a forced landing in French territory, after which he was taken into captivity.

12. & 12a. Uffz. Josef Diem, Albatros D Va

In addition to the Staffel identification markings, the fuselage of the Munich native's aircraft was painted in alternating narrow rings in the colors of his hometown, black and yellow. According to Walter Böning it was the most striking painted aircraft of his Staffel, and he added: *"It was always a good feeling to see his "bee", as his aircraft was nicknamed, or his "Münchner Kindl", by my side during a mission"*.

13. Vzfw. Karl Hopf, Albatros D Va. D.5727/17

After Lt. Hans Böhning's departure from the Staffel, Vzfw. Karl Hopf took over his orphaned Albatros D Va. In addition to the alredy existing white and blue lozenges on the rear fuselage, he added the coat of arms of the "Hochstift Würzburg" (central administration of the archdiocese of Würzburg) on the fuselage of his aircraft, as he was a pupil of the school belonging to the "Hochstift. The school was known to offer a very good free school education for children of non-wealthy parents. It is not the coat of arms of the city of Würzburg as indicated in previous publications. The city crest displays a yellow and red flag with a black background.[3]

14. & 14a. Vzfw. Siegfried Walther, Albatros D Va

As on his Albatros D III (O.A.W.), which only enjoyed a very short service, the fuselage, fin and rudder of this aircraft were painted red. For good measure, a white "W" for "Walther" had also been applied to the fuselage as his personal insignia. The nose and horizontal tail surfaces of this aircraft were painted in Staffel identification marking colors.

15. & 15a. (reverse side) Uffz. Josef Diem, Albatros D Va D.5787/17

In addition to the Staffel identification markings, the second aircraft flown by Uffz. Josef Diem bore the well-known "Münchener Kindl", patterned on the "Oktoberfest Variant" adopted from the city crest of Munich, as his personal marking. Additionally, the aircraft carried a black and yellow fuselage band as a further reference to of his hometown. As seen in this and other profiles of aircraft flown by unit members hailing from Munic, they all found individual ways of referring to their citiy of birth on their personal fighters. This particular aircraft received this paint scheme shortly before the transfer of the Staffel to the German 2nd Army, and was the replacement for his previous "bee".

16. Lt. Hans Hertrich, Albatros D Va

In addition to the unit marking, the fuselage and tail fin of his Albatros D Va was painted "baby-blue". The rudder remained in plain linen fabric complete with "Albatros" company logo. Additionally, the aircraft carried a black "H" with a narrow white border as personal marking, as well as a white-blue-white-blue-white band wrapping around the fuselage in a "Jasta 4"-style manner.

17. Oblt. d. Res. Ernst Angermund, Albatros D Va

Besides the Staffel marking, the aircraft had been painted with his personal insignia around the fuselage; a red band edged in white. These markings were a reference to his parent unit, the Bavarian 9th Infantry-Regiment, deploying with the Bavarian 4th Infantry-Regiment at the beginning of the war. Both units had the usual red cap bands of the infantry. The rudder of this aircraft remained in five-color fabric.

18. Vzfw. Karl Koller, Albatros D Va D.7221/17

The aircraft was flown by Vzfw. Koller when he had to make an emergency landing in British lines following aerial combat on May 25, 1918. According to the British investigation report issued after the aircraft was captured (File No. G/3Brigade/7), the front part of the fuselage up to and including the cockpit were yellowish wood, whereas the rear part of the fuselage was painted brown. A close examination of the two available photos shows that the fuselage at the cockpit was also painted with a flame designed edge. The white "M" on both sides of the fuselage had been painted on a light-colored band, very likely in Bavarian blue. Additionally, the fin and rudder were not brown, but painted white. The aircraft's nose and horizontal tail surfaces were painted in Staffel markings. This is another case proving that the descriptions of the colors of captured German airplanes in the British G-Reports were not entirely accurate, as the fuselage band and the letter "M", as well as the "flaming edges", are not mentioned in the report.[4]

19. Vzfw. Karolus Bärenfänger & Lt. d. Res. Max Schick, Albatros D Va D.5765/17 (Interpretation)

This Albatros D Va was Lt. Karolus Bärenfänger's second aircraft in Jagdstaffel 76. It was flown by Lt. Walter Böning on May 31, 1918 when he was injured in aerial battle on that day. After Lt. Bärenfänger left the unit on June 21, 1918, the aircraft was taken over by Lt. Max Schick. He was shot down while flying this aircraft just two days later, on June 23, 1918. Incredibly, 95 years after this event, major components of the aircraft wreckage were discovered in a swamp near the village of Boussicourt. (5) Unfortunately, no photo of this particular aircraft in its final guise could be found. However, a close investigation of the recovered components of the aircraft indicates that it was painted in the same way as the previous Albatros D Va of Lt. Bärenfänger, see Profile 11 & 11a. In June 1918, the aircraft was certainly marked with the Balkenkreuz instead of the Iron Cross. This was the second (former) personal aircraft of Lt. Bärenfänger to be lost with another pilot at the controls over enemy territory.

20. Uffz. Josef Bauer, Albatros D Va

Besides having received the usual Staffel marking on the nose and horizontal tailplane, Uffz. Josef Bauer's Albatros D Va was marked with a green and brown speckled fuselage. This was a rare case of an attempt at applying a camouflage scheme to an aircraft at Jagdstaffel 76. However, this was somewhat offset by the blue painted wheel hubs. The wings of this aircraft were covered in five-color fabric, and the crosses had been converted to the thick, "late"-style Balkenkreuz marking when the plane was photographed.

21. Uffz. Josef Bauer Albatros D Va D.7544/17

This very late-production Albatros D Va was not marked with any kind of personal insignia, indicating that it likely served as a reserve aircraft. The national markings on the aircraft had undergone several modifications, indicating that the airframe, in spite of its reserve status, had still seen long service before it crashed. Uffz. Josef Bauer used it on the units´ last flight from Bignicourt airfield on August 18, 1918 and crash-landed after being attacked by several Allied aircraft. He emerged from the wreck unharmed.

By the time of the crash, the crosses had been converted to the thick, "late"-style Balkenkreuz marking. We have shown the "reduced" parts of the fuselage and rudder crosses as being overpainted in blue color, although this is unconfirmed.

Pfalz D XII Factory Finish

It is noticeable that on a number of Pfalz D XIIs the fuselage appears somewhat darker duller and less reflective than the fuselages of the silver-grey Pfalz D III/IIIas. Documents discovered by Colin Owers in the Australian War Memorial in Canberra and Greg VanWangarden lead to the conclusion that the fuselages of the Pfalz D XIIs painted in a single color were finished with a grey-blue fuselage. Most likely, this was the same grey-blue color found at the later D.XIIs that were finished in the multi-colored camouflaged fuselage schemes. The exact shade of this blue-grey color is not known, unfortunately. The colour profiles were prepared with an interpretation of this shade resembling the blue-grey version of the later RLM 78 color.

22. Staffel marking Pfalz D XII (Top view tail)
On the new Pfalz D XII fighters, the top and bottom surfaces of the horizontal tail planes were painted in Bavarian blue with a white lengthwise stripe centred on each side, just as they were on the previous Albatros fighters. However, the blue and white nose markings used previously were not carried over to the Pfalz, which retained their noses in factory finish.

23. Gefr. Julius Geyer, Pfalz D XII
In addition to the unit markings, the aircraft was marked with the black and white ribbon of the Iron Cross II. Class as personal insignia painted around the rear fuselage. Available photographs indicate that on the Pfalz D XIIs operated by Jagdstaffel 76 the personal markings were applied to the rear fuselages, behind the fuselage cross. As a result, these markings covered the factory-applied military numbers on the fuselages of the D XIIs.[6]

24. Pfalz D XII, unknown pilot (Interpretation)
In addition to the Staffel identification markings, this Pfalz D XII aircraft had a horizontal striped tail. The color of the stripes was interpreted to be red, a common choice of personal colour for pilots coming from the infantry. It is also possible that the colors were black-white-red. According to allied

25. Pfalz D XII, unknown pilot (Interpretation)
Pfalz D XII was marked with a fuselage band presumed to be white-blue-white, a variant of the Bavarian colors.

26. & 26a. (Top-view) Uffz. Josef Bauer, Pfalz D XII
The aircraft had a broad white fuselag band, positioned behind the Balkenkreuz, as well as a similar broad white stripe on the upper wing, the identification of a "Ketten-leader".[7]

27. Staffel marking Fokker D VII (Top view horizontal tailplanes)
As on the previous Pfalz D XII fighters, the top and bottom surfaces of the horizontal tail planes of the Fokker D VII were painted in Bavarian blue, with a white lengthwise stripe centered on each side. The noses of the aircraft remained in factory finish. The Fokker- and Albatros-built D VIIs operated by Jasta 76 were apparently covered in four-color aircraft fabric, while most O.A.W.-built D VIIs (with the exception of Geyer´s 6381/18) seem to have been covered in five-color fabric.

28. & 28a. (Top-view) Lt. Eberhard Grosse, Fokker D VII
In addition to Staffel markings applied to the horizontal tail surfaces, alternating white-yellow-white-yellow-white fuselage bands were painted onto the rear fuselage of his aircraft. These were the colors of the Prussian province of Silesia, where he was born, and they were his personal markings. All indications are that he brought along this Fokker from his previous unit, Jagdstaffel 72. Being a Saxon unit, the noses of the Fokkers operated by this Staffel are thought to have been painted in a slightly more yellowish green, which contrasted with the darker, factory-applied "Fokker" green applied to the metal components. Since the aircraft is not seen in the lineup photo taken in October 1918 at Habsheim, it is thought that Grosse took the plane with him when he returned to Jagdstaffel 72 on October 7. The military number of this Fokker-built D VII is not known, but it was a plane from the 4250/18 – 4449/18 production batch, and it was covered in four-color fabric at the factory.

29. & 29a. (Top-view) Oblt. Amandus Rostock, Fokker D VII (O.A.W.) 6372/18
The aircraft had the tail in a dark colour bordered by a white stripe. The dark tail was interpreted as red, because Oblt. Rostock used the same colour as personal marking as a member of Jagdstaffel 77. As Otto Fuchs remarked: "He was a proud infantry officer".[8] On the centre-section of the upper wing a broad blue stripe was painted identifying his aircraft as the one of the Staffelführer. This late-production O.A.W.-built D VII was covered in five-color fabric.

30. Uffz. Julius Geyer, Fokker D VII (O.A.W) 6381/18
Similar to his previous Pfalz D XII, the aircraft bore the two-colored band of the Iron Cross Class II around the fuselage, just behind the cockpit. The horizontal tailplanes bore the Staffel markings, with the blue paint being applied below the tailplane as well. The fuselage of this plane was covered in four-color fabric.

31 & 31a. Uffz. Josef Bauer, Fokker D VII (O.A.W)
In an original photograph in Josef Bauer's possession,

several Fokker D VIIs are shown lined-up in front of the hangars at Habsheim; one of the aircraft had a light band painted on it. On the back of the original photograph, an arrow points to that particular aircraft. According to a statement made by Josef Bauer's daughter, this was the aircraft flown by her father. The band in the photograph appears very light and was very likely white. The airframe was covered in five-color fabric, and the horizontal tailplane carried the unit marking.

32. Lt. Karl Dumler, Fokker D VII (Alb.)

This Albatros-built aircraft was marked with the personal insignia of a wave bar painted along the fuselage sides. The color of this marking was interpreted as white. The photo of Dumler sitting in this aircraft indicates that this D VII was covered in four-color fabric.

33. Fokker D VII, unknown pilot, (Interpretation)

This Fokker-built aircraft carried a white-dark-white band around the fuselage. It was interpreted white-black-white fuselage band, if only to allow for better differentiation with the aircraft shown in the next profile. The horizontal tail surfaces carried the unit markings. Like the D VII flown by Grosse, this was a Fokker-built aircraft covered in four-color fabric.

34 & 34a. Lt. Karl Neu, Fokker D VII (O.A.W.)

This aircraft was marked with a red band edged in white, which was painted onto the fuselage between the cockpit and the cross. This was a reference to his former unit, the 9th Bavarian Infantry-Regiment. The center section of the upper wing was adorned with a blue "N". Similar markings can barely be made out on several of the D VIIs seen in the lineup photo taken at Habsheim. It is impossible to determine if this was done as a "Kette"

35. & 35a. Lt. Walter Curtius, Fokker D VII (O.A.W.) 6481/18 (Interpretation)

Several photos of this aircraft were taken following the armistice, after it was handed over to the Allies. The military number of the plane can barely be deciphered on the left side of the fin and rudder in one of these photos. The bottom of the radiator nose plate of this aircraft was painted with a "sunburst" insignia. The color of the sun and sunbeams marking was interpreted as red and the backdrop of the sunbeams blue. According to our interpretation the fuselage band was also red-blue-red and in this case would represent the colours of the uniform of his parent unit, the Westfälisches Ulanen-Regiment no. 5, which that had blue uniforms with red lapels and red cap-bands. The airframe was covered in five-color fabric, and the horizontal tailplane carried the unit marking.

36. & 36a. Lt. Maximilian Veith, Fokker D VII (Alb.) 5341/18 (interpretation)

This aircraft appears in the lineup photo at Habsheim and was also photographed several times after the war in Allied hands. All of these photos show at least one replacement component on this Albatros-built D VII: The rudder was taken from a Fokker-built D VII. Unfortunately, the Fokker works number, applied to the right bottom of the rudder, is not clear enough to be deciphered. The fuselage and lower wing were covered in four-color fabric, with rib tapes from the same material. The upper wing is also covered in four-color fabric, but the rib tapes are either light blue or plain linen, we have chosen the former color for the profile.

His aircraft carried a horizontal stripe on the fuselage, interpreted as red for the profile, and a "V" on the upper wing. The color of the "V" is not known and interpreted as blue.

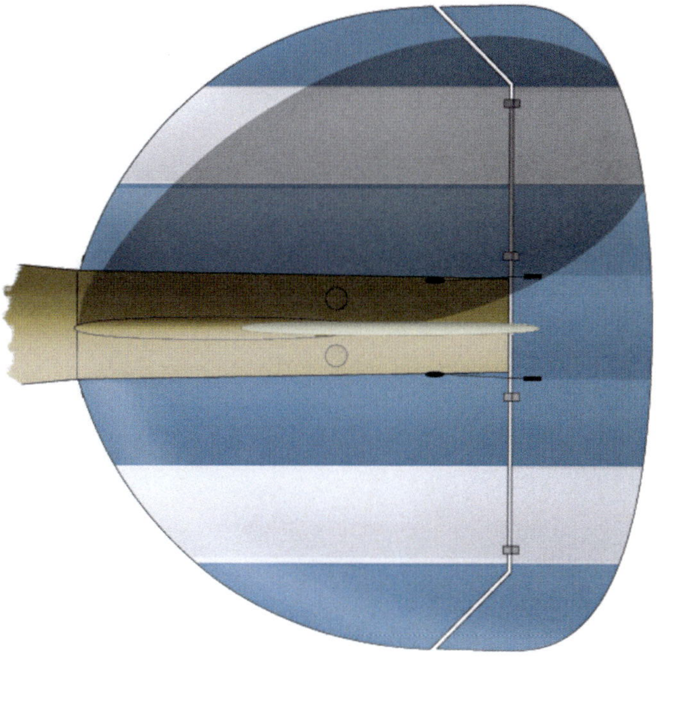

1. Staffel marking of Albatros D III (O.A.W.) and D Va aircraft.

2. Lt.d.Res. Walter Böning, Albatros D III (O.A.W.)

3. Lt.d.Res. Franz Xaver Danhuber,
Albatros D III (O.A.W.)

4. Vzfw. Paul Wagner, Albatros D III (O.A.W.)

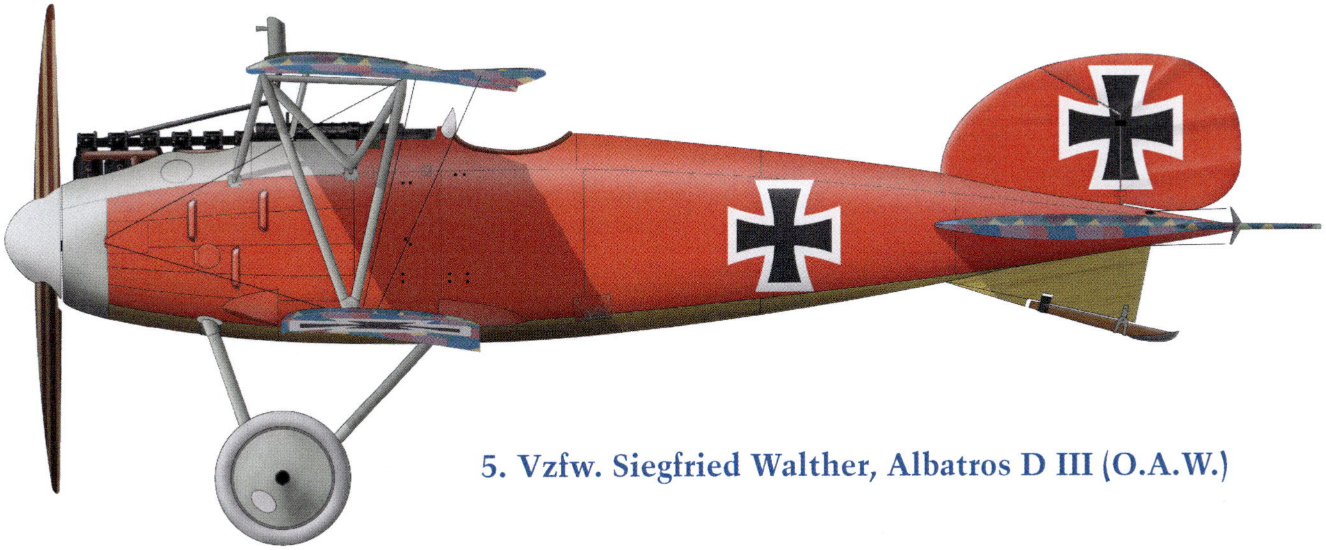

5. Vzfw. Siegfried Walther, Albatros D III (O.A.W.)

6. Oblt. Erwin Wenig, Albatros D III (O.A.W)

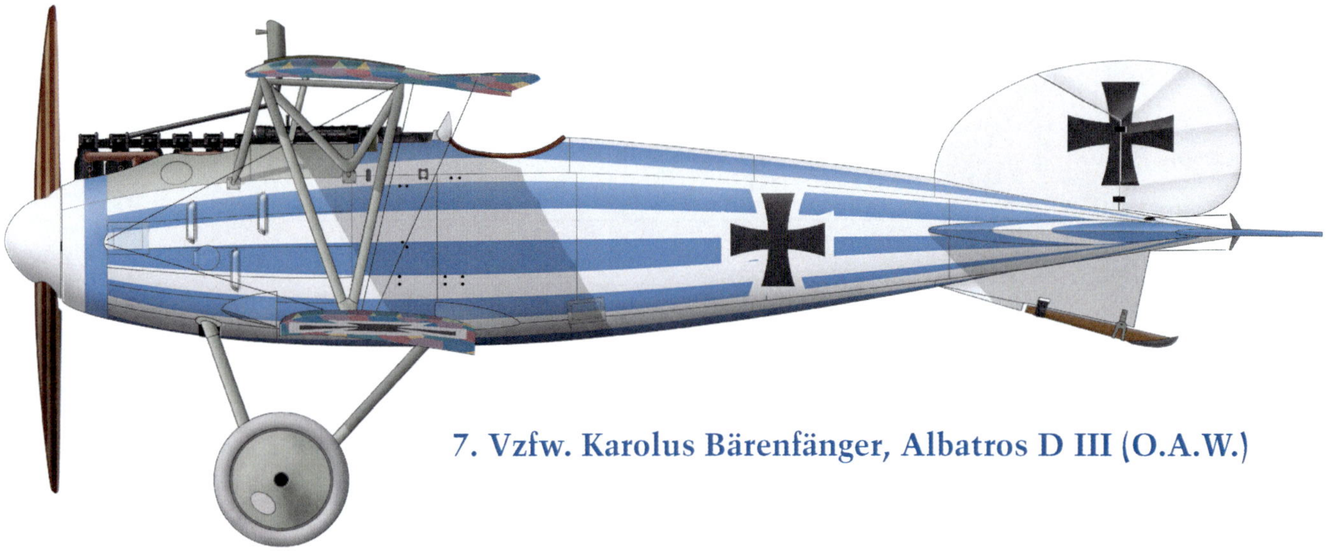

7. Vzfw. Karolus Bärenfänger, Albatros D III (O.A.W.)

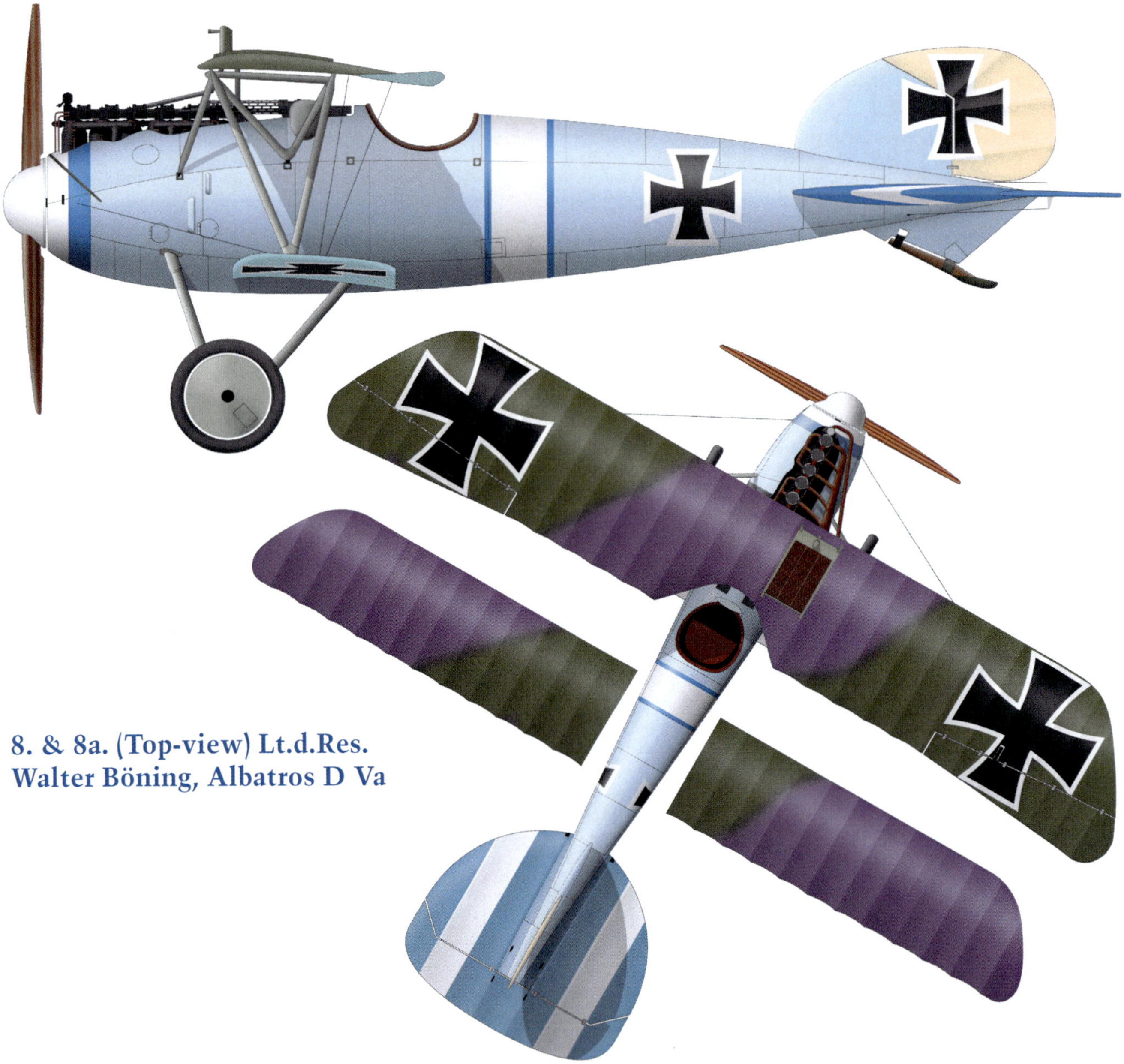

8. & 8a. (Top-view) Lt.d.Res. Walter Böning, Albatros D Va

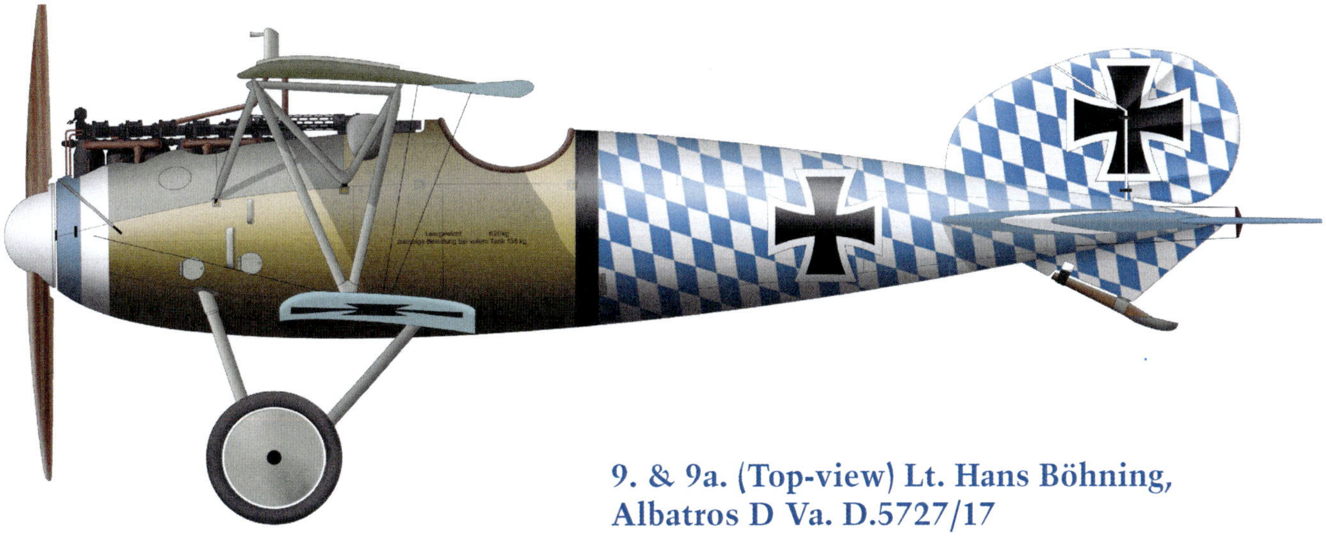

9. & 9a. (Top-view) Lt. Hans Böhning, Albatros D Va. D.5727/17

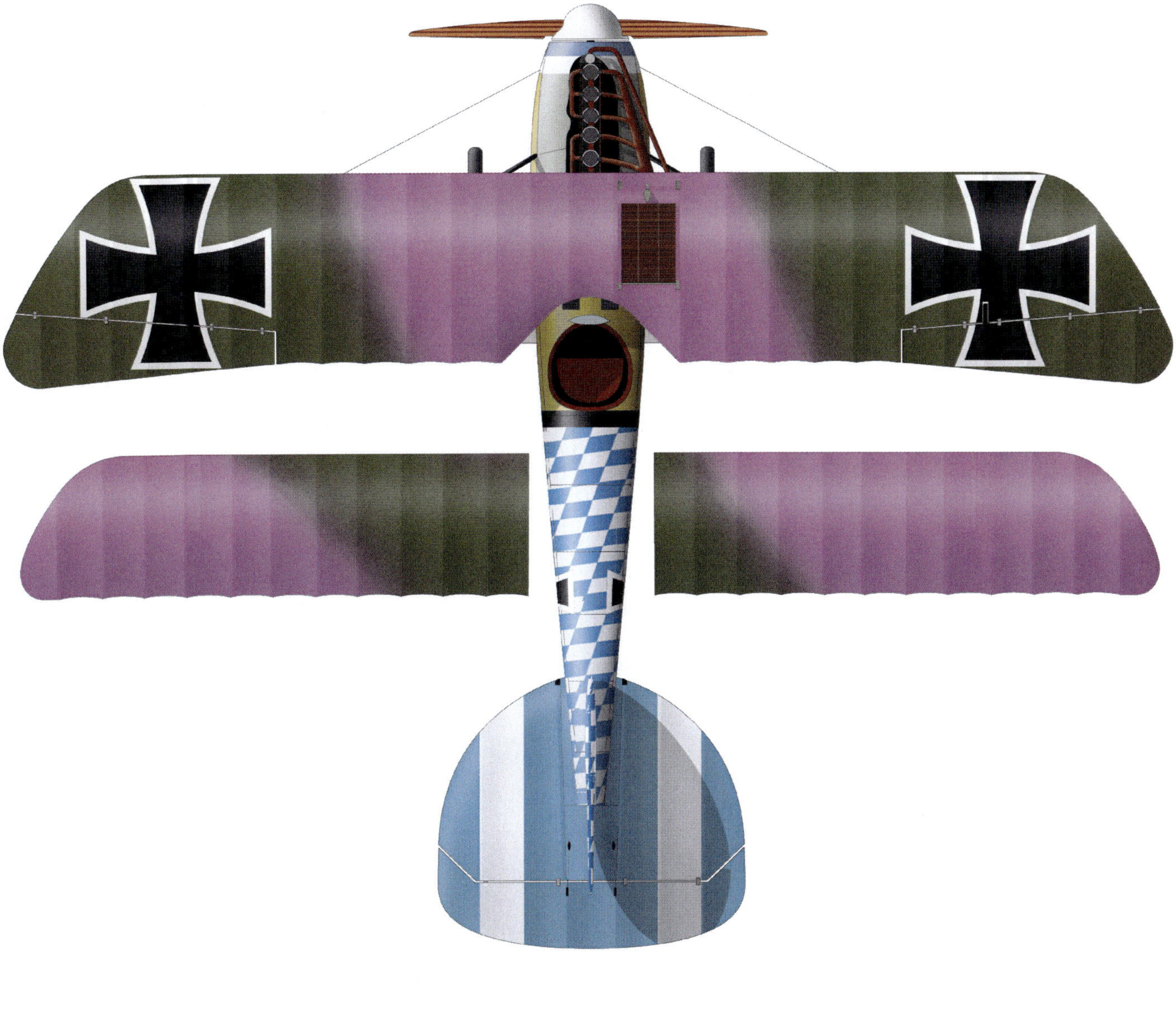

10. Lt.d.Res. Wilhelm Kolb, Albatros D Va D.5294/17

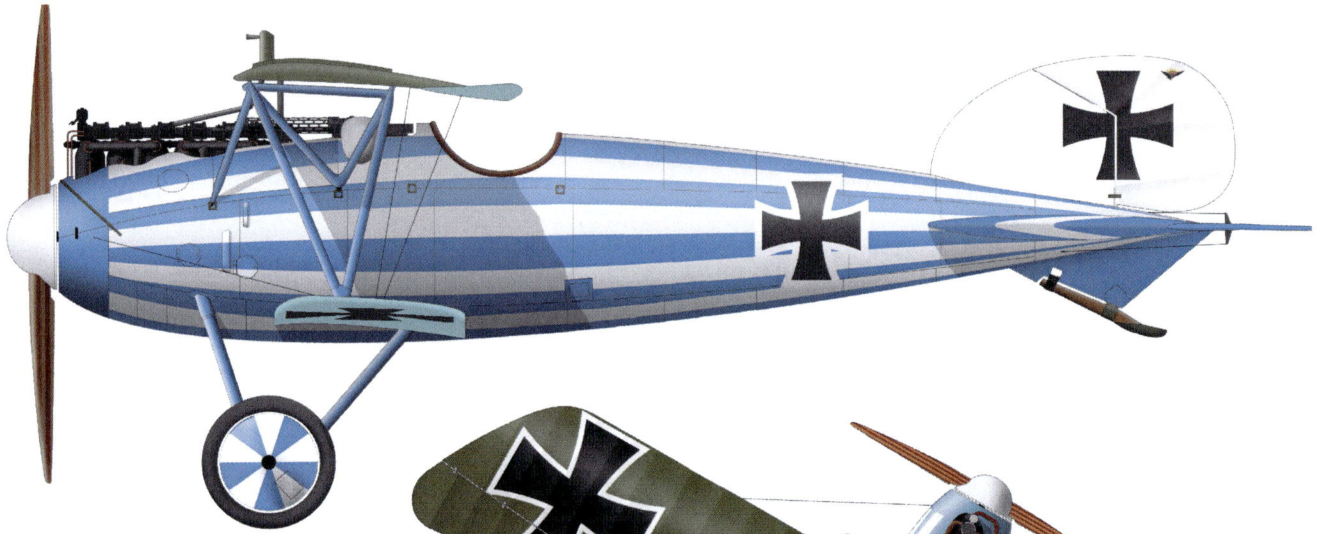

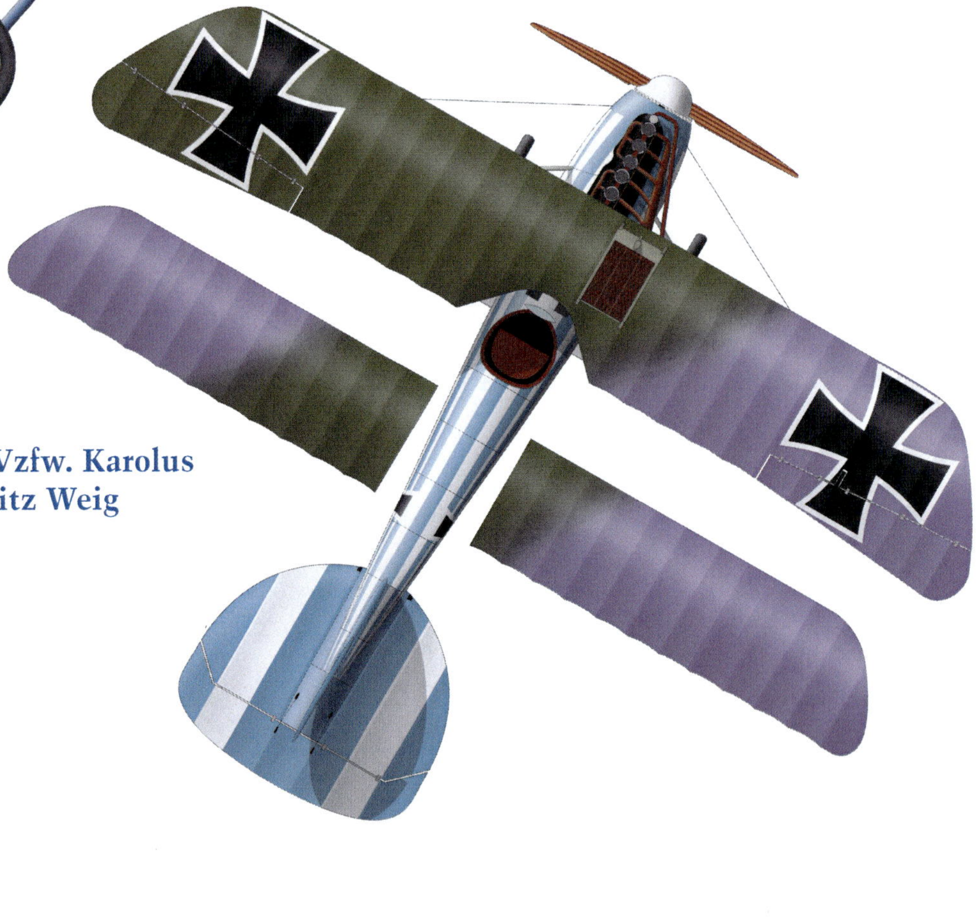

11. & 11a. (Top-view) Vzfw. Karolus Bärenfänger & Lt. Moritz Weig Albatros D Va

12. & 12a. (Top-view) Uffz. Josef Diem, Albatros D Va

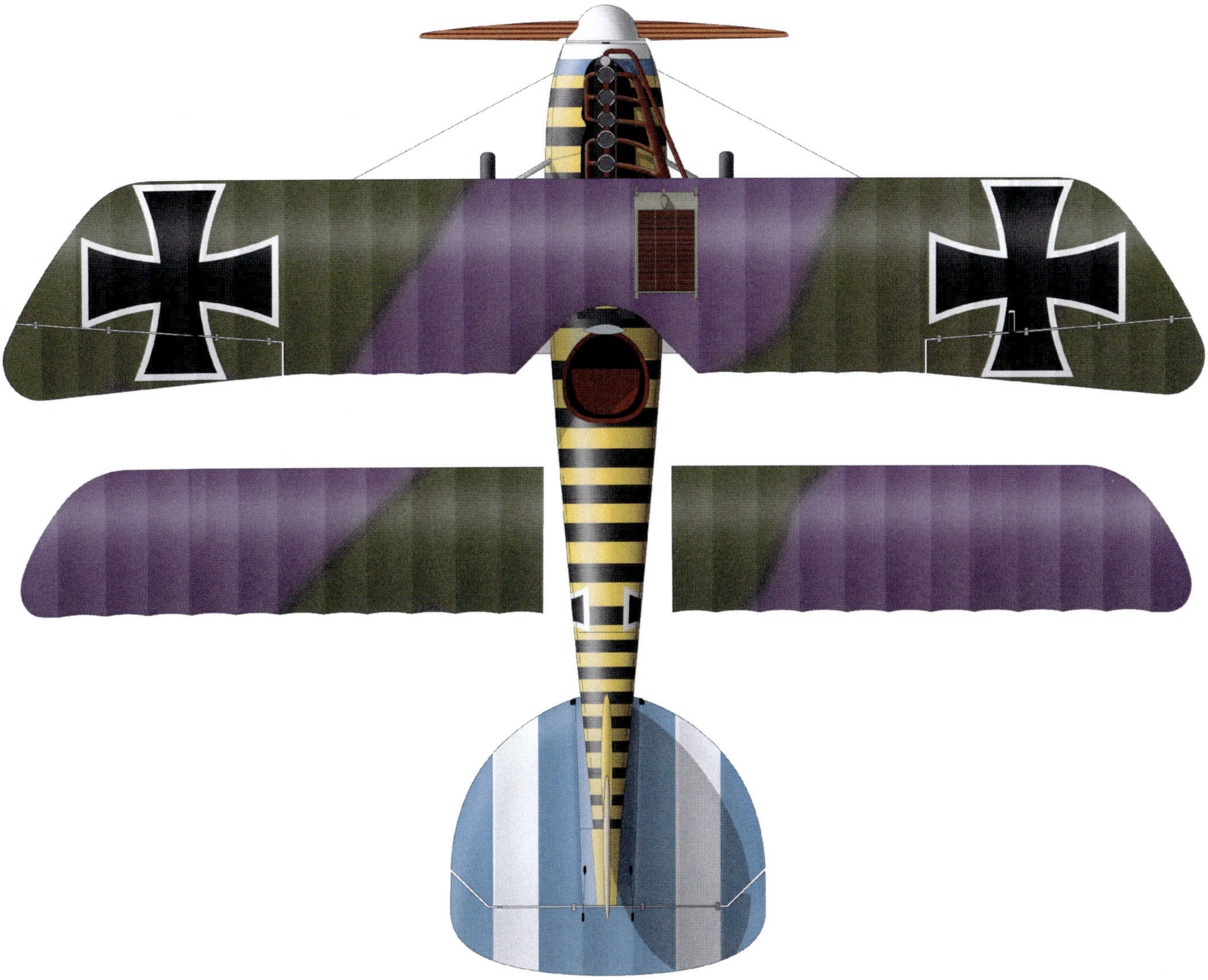

13. Vzfw. Karl Hopf, Albatros D Va. D.5727/17

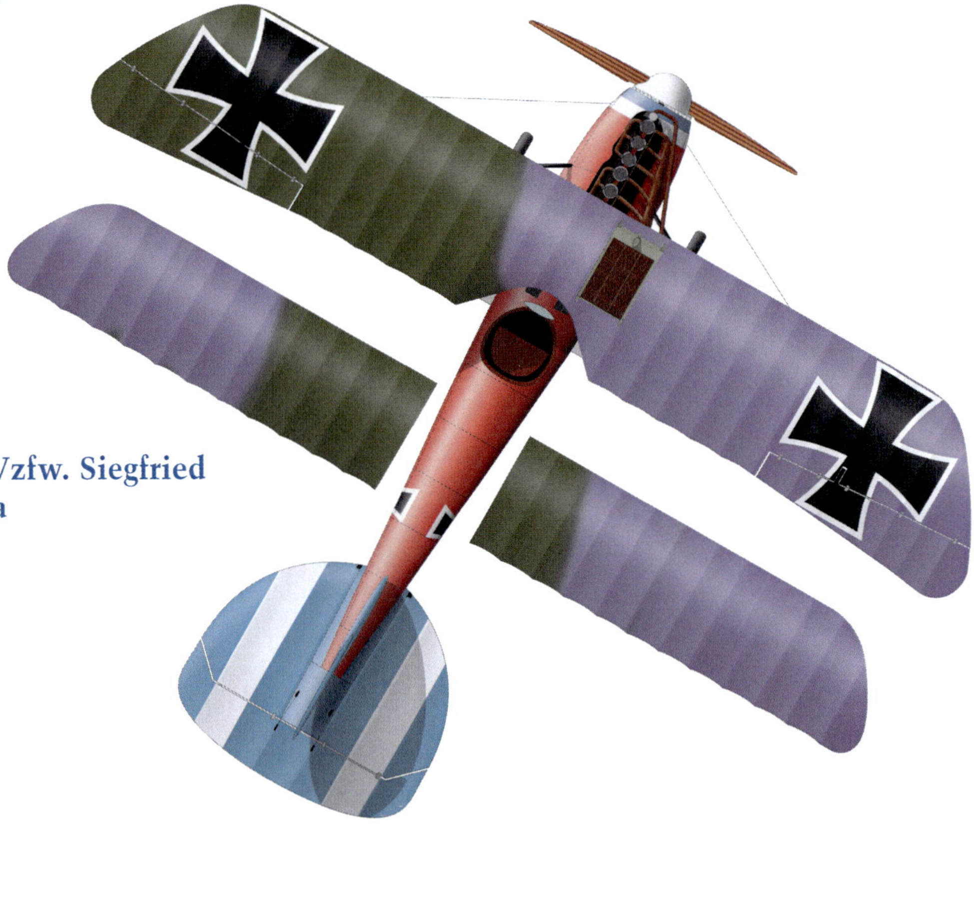

14. & 14a. (Top view) Vzfw. Siegfried Walther, Albatros D Va

15. & 15a. (reverse side) Uffz. Josef Diem,
Albatros D Va D.5787/17

16. Lt. Hans Hertrich, Albatros D Va

17. Oblt. d. Res. Ernst Angermund, Albatros D Va

18. Vzfw. Karl Koller, Albatros D Va D.7221/17

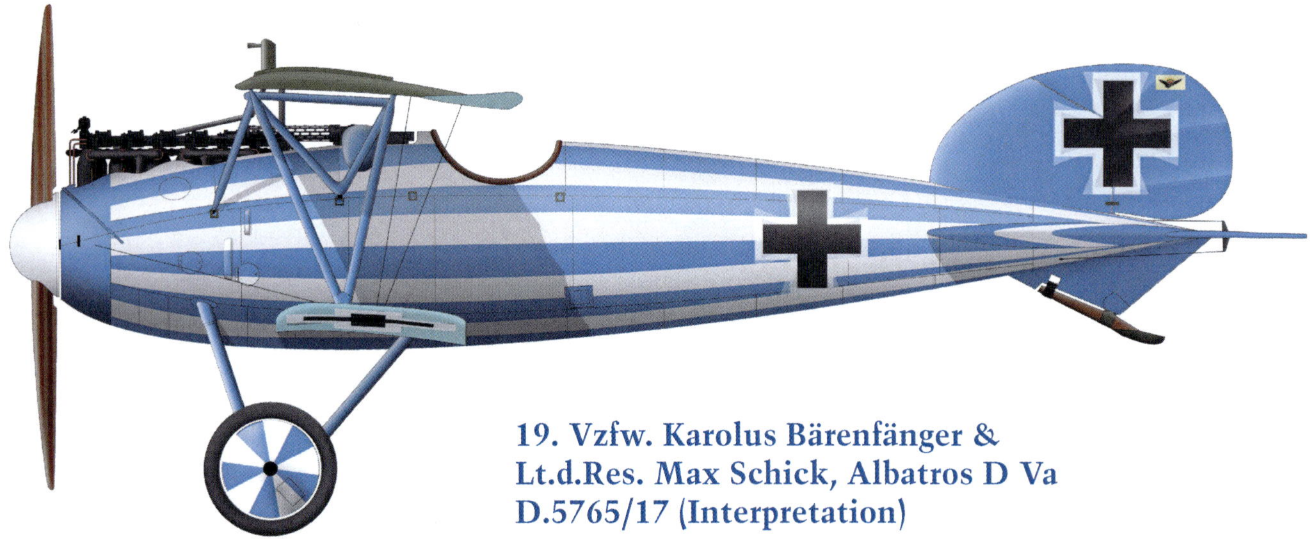

19. Vzfw. Karolus Bärenfänger & Lt.d.Res. Max Schick, Albatros D Va D.5765/17 (Interpretation)

20. Uffz. Josef Bauer, Albatros D Va

21. Uffz. Josef Bauer Albatros D Va D.7544/17

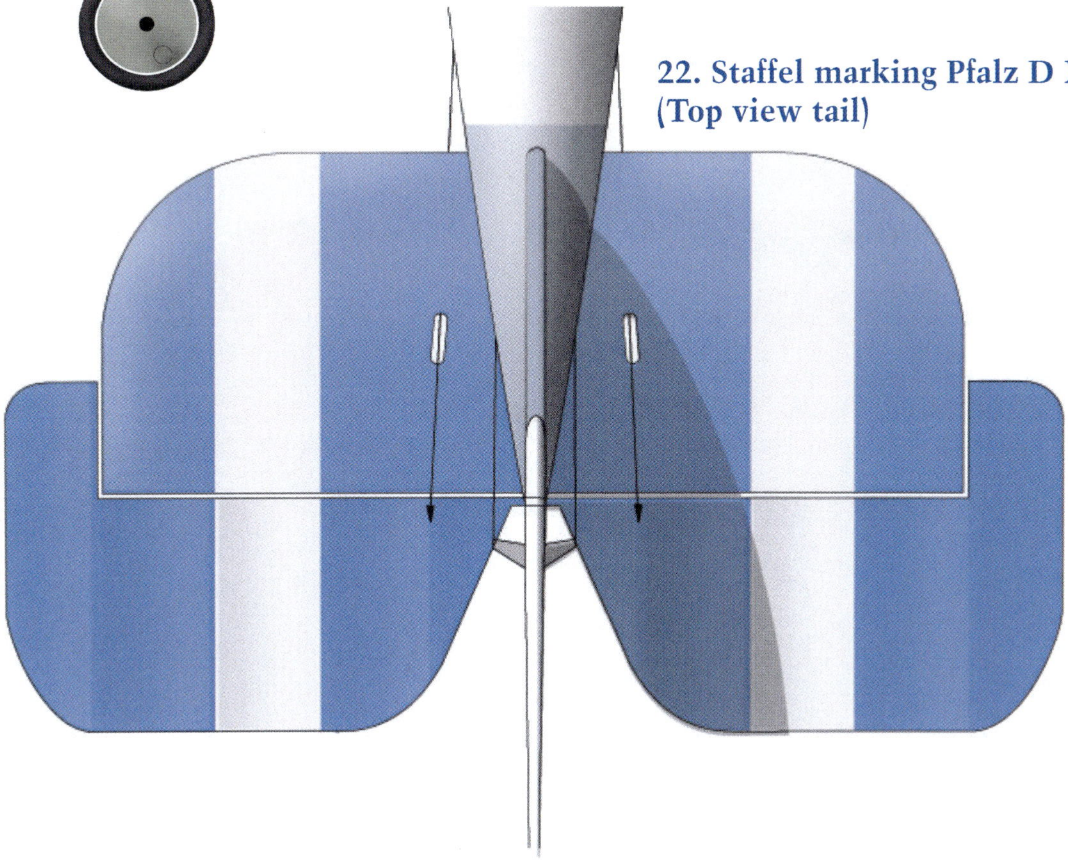

22. Staffel marking Pfalz D XII
(Top view tail)

23. Gefr. Julius Geyer, Pfalz D XII

24. Pfalz D XII, unknown pilot (Interpretation)

25. Pfalz D XII, unknown pilot (Interpretation)

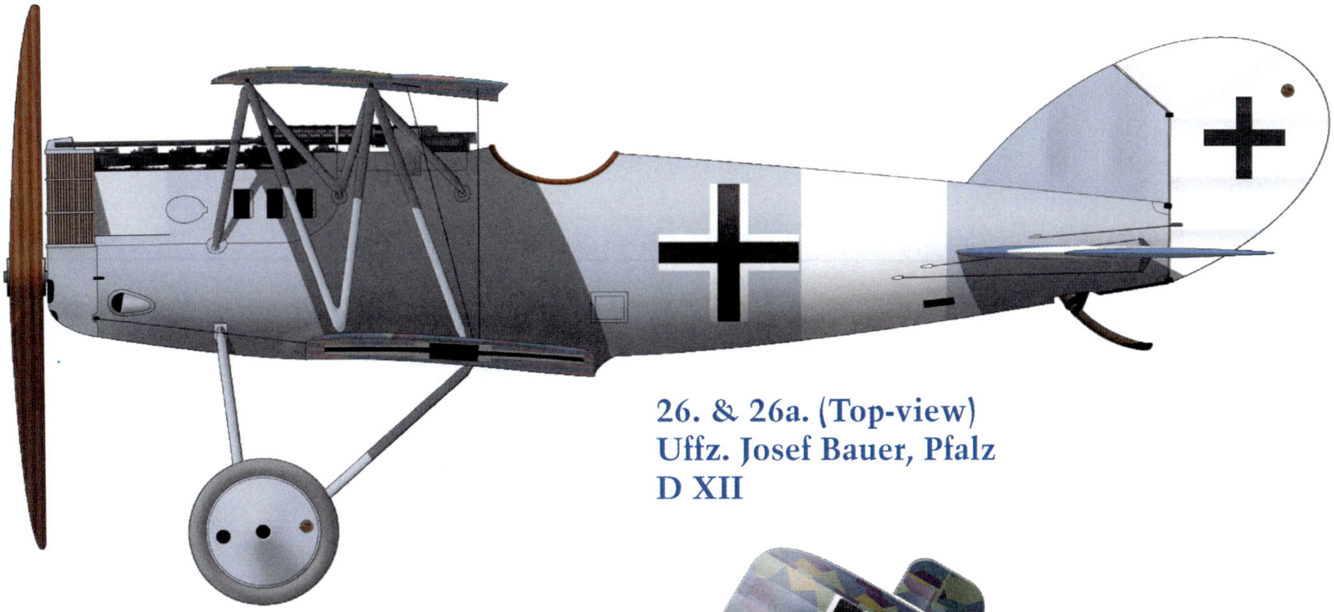

26. & 26a. (Top-view)
Uffz. Josef Bauer, Pfalz
D XII

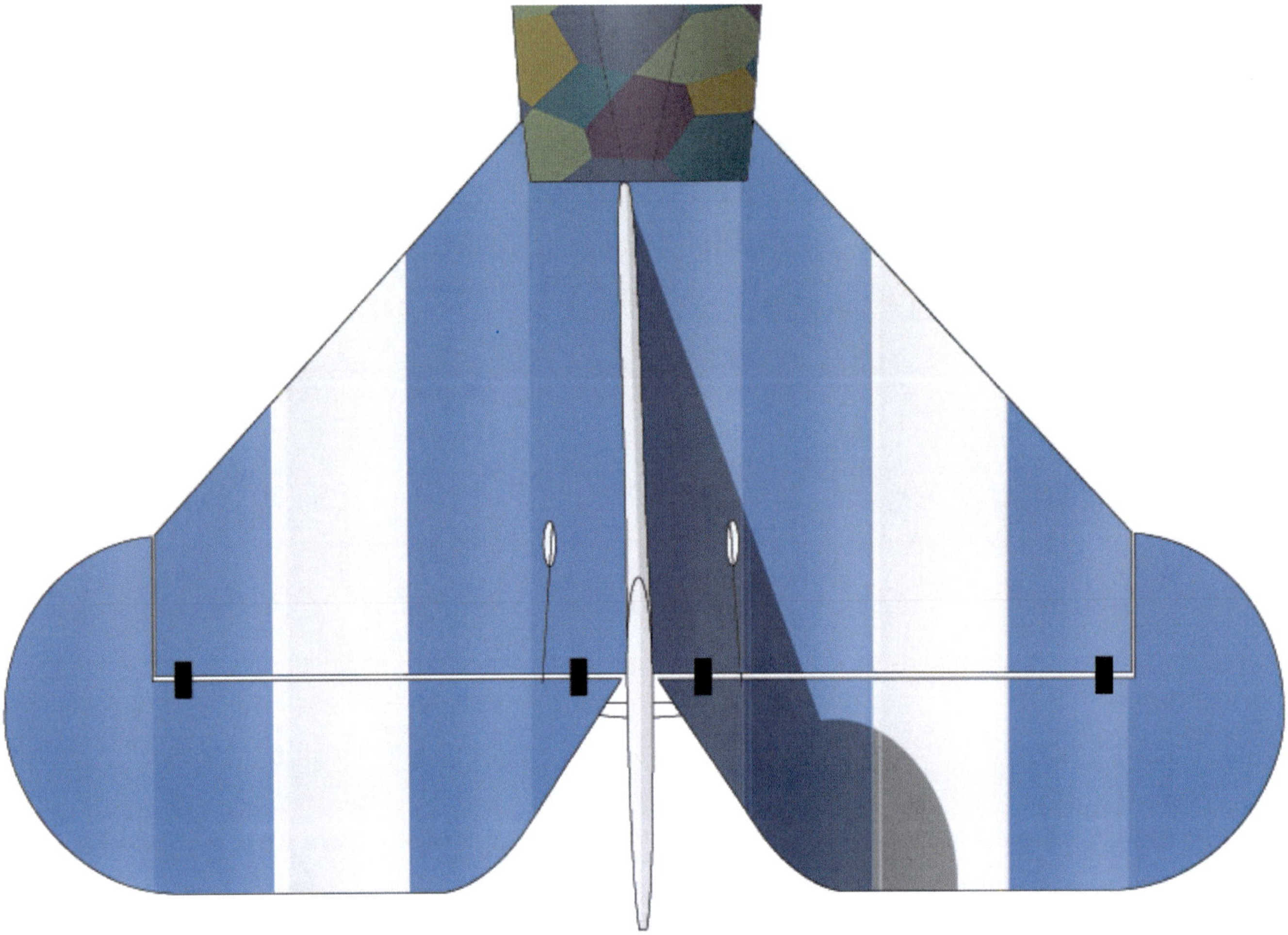

27. Staffel marking Fokker D VII (Top view horizontal tailplanes)

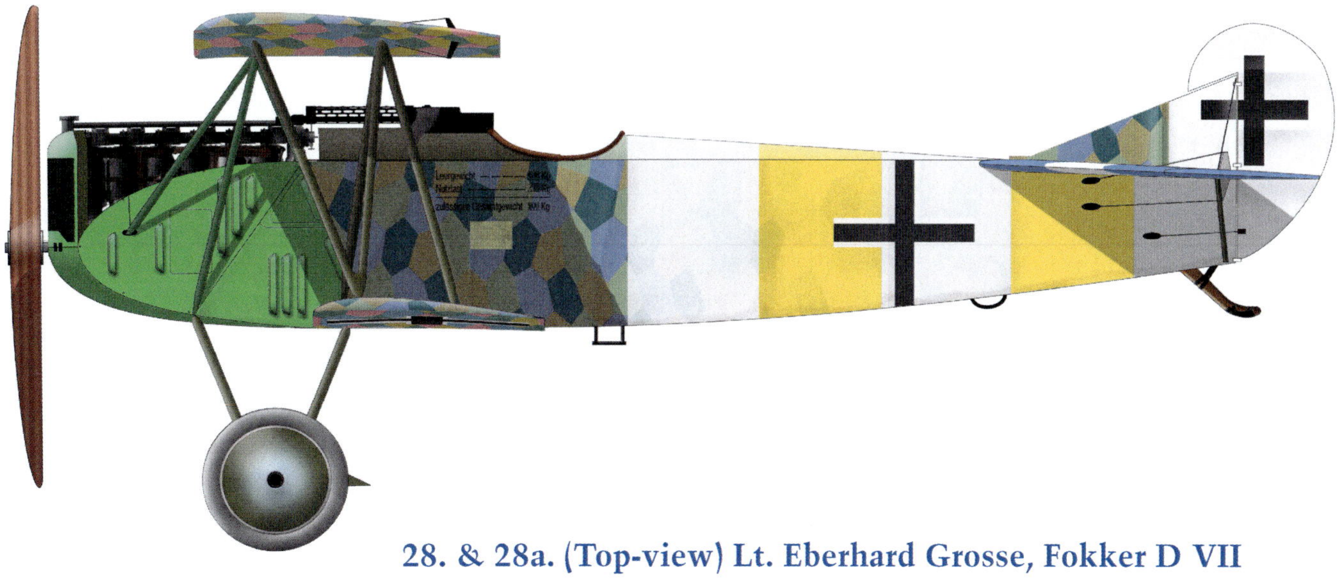

28. & 28a. (Top-view) Lt. Eberhard Grosse, Fokker D VII

29. & 29a. (Top-view) Oblt. Amandus Rostock, Fokker D VII (O.A.W.) 6372/18

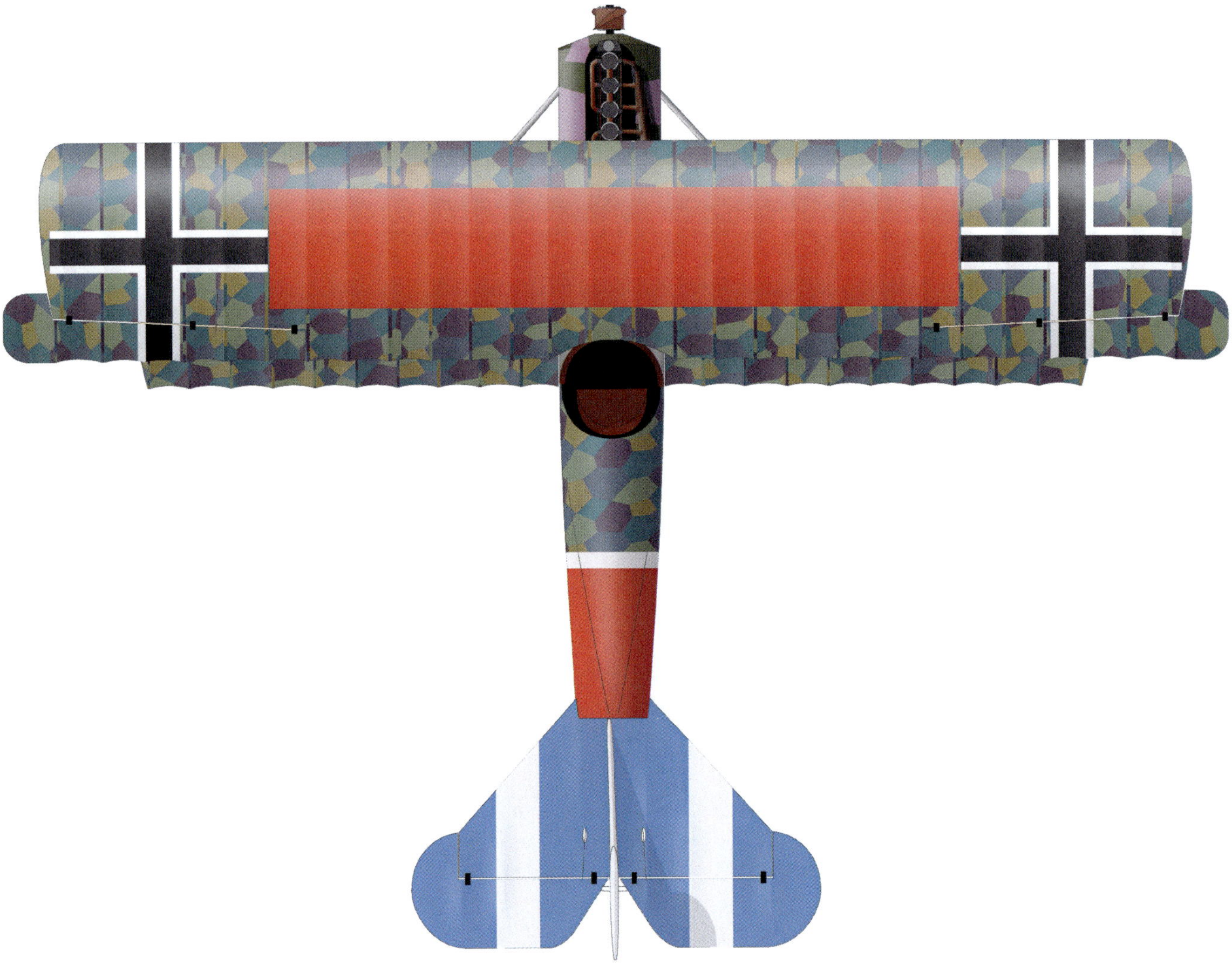

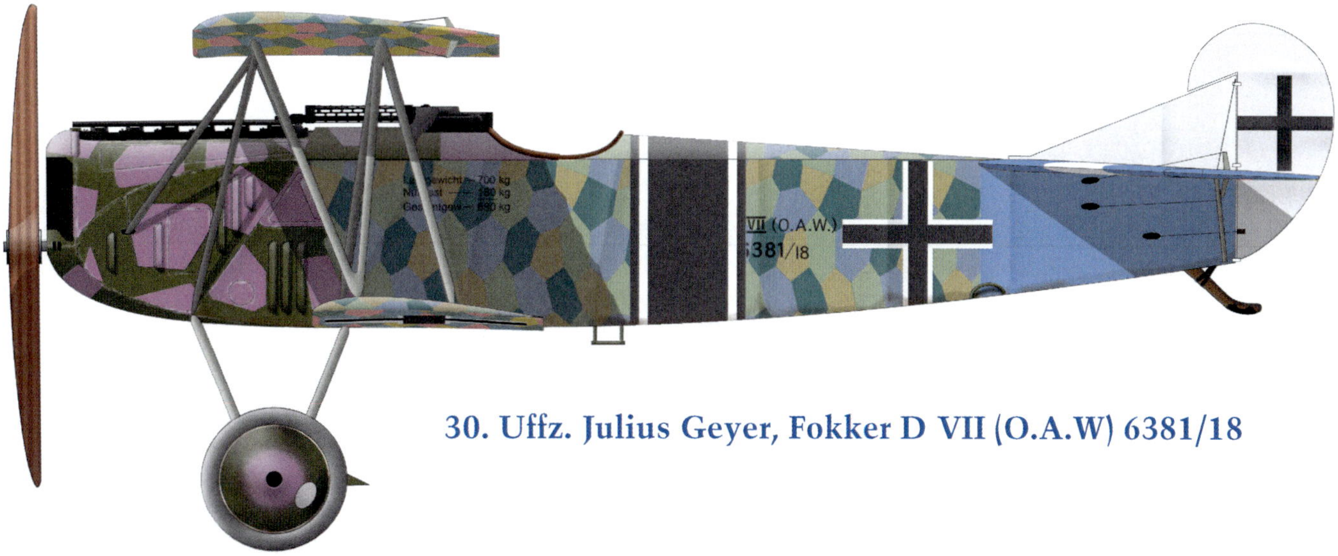

30. Uffz. Julius Geyer, Fokker D VII (O.A.W) 6381/18

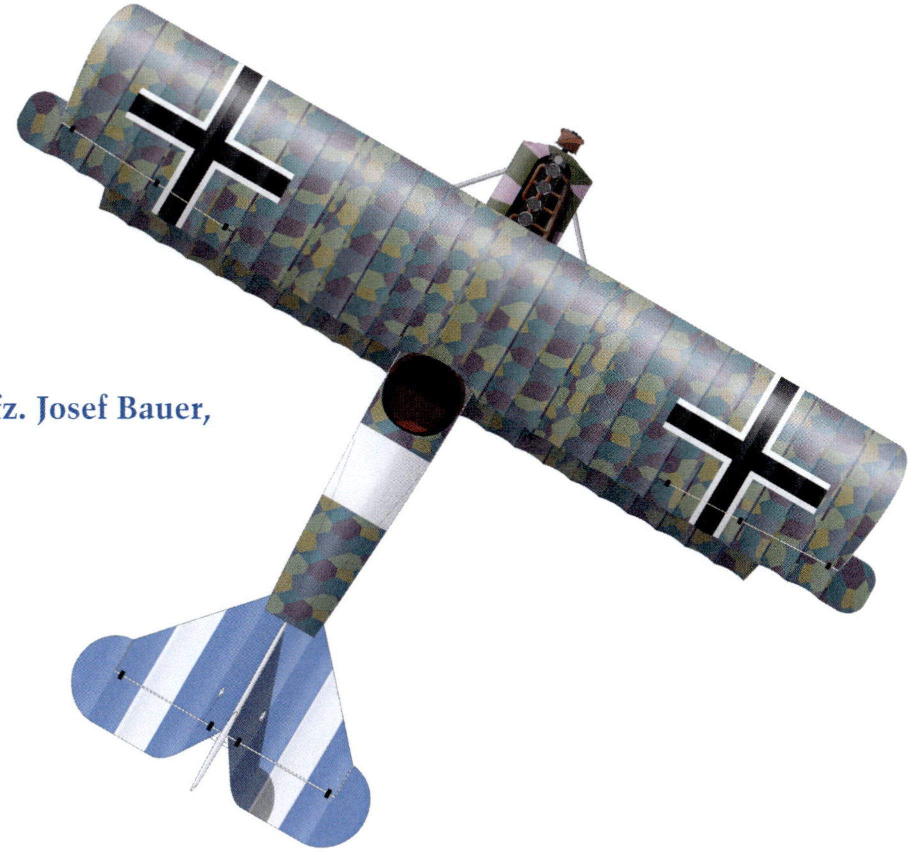

31 & 31a. (Top-view) Uffz. Josef Bauer,
Fokker D VII (O.A.W)

32. Lt. Karl Dumler, Fokker D VII (Alb.)

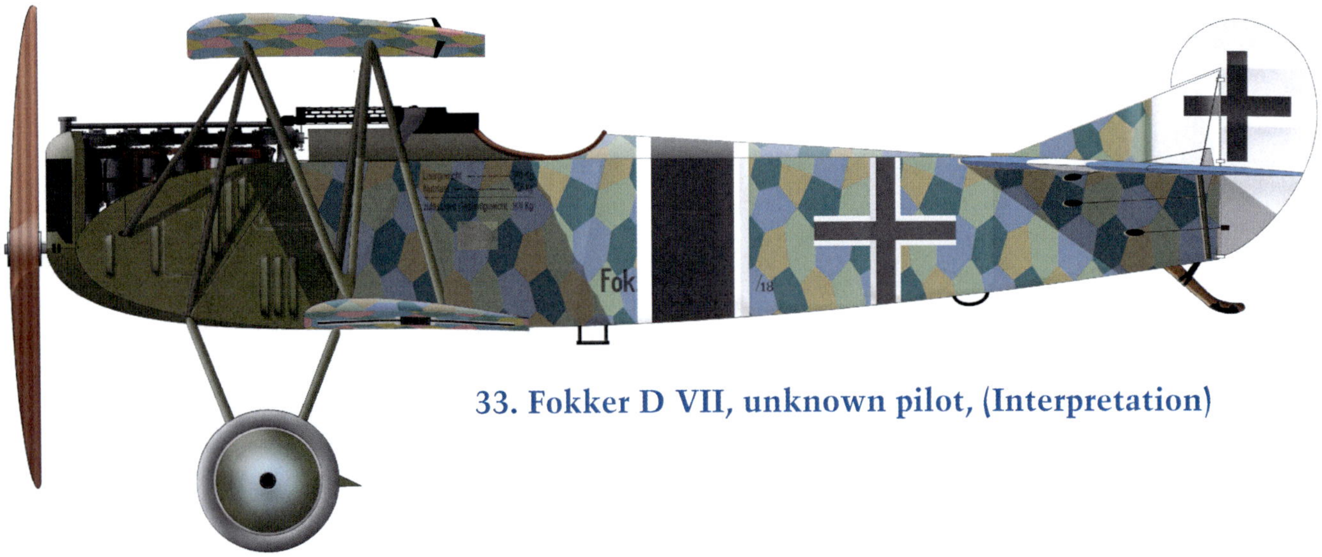

33. Fokker D VII, unknown pilot, (Interpretation)

34 & 34a. (Top-view) Lt. Karl Neu,
Fokker D VII (O.A.W.)

35. & 35a. (Front-view) Lt. Walter Curtius,
Fokker D VII (O.A.W.) 6481/18 (Interpretation)

36. & 36a. (Top-view) Lt. Maximilian Veith, Fokker D VII (Alb.) 5341/18 (interpretation)

8 Royal Bavarian Jagdstaffel 76

Flying Personnel

Rank	Last Name	First Name	Arrival	Departure	Remarks
1. Staffelführer					
Lt.d.Res.	Böning	Walter	05.10.1917	31.05.1918	WIA
Lt.	Schmid	Ludwig	20.06.1918	16.07.1918	WIA
Oblt.d.Res.	Rostock	Amandus	07.10.1918	EOW	
2. Offizier zur besonderen Verwendung					
Lt.d.Res.	Vollert	Clemens	18.11.1917	16.02.1918	Jasta 78 Offz. zbV
Lt.d.Res.	Vollert	Clemens	25.02.1918	EOW	
3. Acting Staffelführer					
Oblt.	Angermund	Walter	29.04.1918	23.05.1918	WIA
Lt.d.Res.	Schick	Max	01.06.1918	23.06.1918	POW
Lt.	Grosse	Eberhard	08.08.1918	07.10.1918	Jasta 72
4. Pilots					
Oblt.	Angermund	Walter	03.04.1918	23.05.1918	WIA
Uffz.	Artmann	Ludwig	14.04.1918	12.07.1918	KIA
Lt.d.Res.	Bärenfänger	Karolus	15.11.1917	21.06.1918	FEA 1b
Uffz.	Bauer	Josef	23.07.1918	EOW	
Lt.d.Res.	Böhning	Hans	06.11.1917	23.02.1918	Staffelführer Jasta 79
Uffz.	Brunner	Johann	05.07.1918	EOW	
Lt.d.Res.	Curtius	Walter	27.06.1918	EOW	
Lt.d.Res.	Danhuber	Franz Xaver	15.11.1917	13.12.1917	Staffelführer Jasta 79
Uffz.	Diem	Josef	13.02.1918	EOW	
Lt.d.L.	Dumler	Karl	20.06.1918	EOW	
Lt.d.Res.	Emmerich	Richard	25.03.1918	11.04.1918	KIA
Uffz.	Geyer	Julius	23.07.1918	EOW	
Uffz.	Hainzl	Rupert	29.07.1918	EOW	
Uffz.	Hepperlin	Edwin	30.07.1918	EOW	
Lt.d.Res.	Hertrich	Hans	16.02.1918	06.04.1918	AFP 2 commanded and transferred to AFP 8
			27.07.1918	04.09.1918	FEA 1b
Uffz.	Hopf	Karl	12.12.1917	20.04.1918	AFP 2 later Jasta 80
Vzfw.	Hüttner	Franz	29.02.1918	06.04.1918	AFP 2
Lt.	Kolb	Wilhelm	05.11.1917	12.12.1917	KIC
Vzfw.	Koller	Karl	07.04.1918	25.05.1918	POW

Vzfw.	Markert	Georg	04.05.1918	31.05.1918	KIA
Lt.d.Res.	Martin	Franz	04.05.1918	28.05.1918	WIA
Lt.d.Res.	Martin	Franz	05.11.1918	EOW	
Lt.	Neu	Karl	23.04.1918	16.07.1918	WIA
	Neu	Karl	12.08.1918	EOW	
Lt.d.Res.	Obermeier	Hermann	23.04.1918	11.05.1918	Hospital
	Obermeier	Hermann	22.07.1918	EOW	
Uffz.	Reinhard	Josef	05.07.1918	EOW	
Lt.d.R.	Schick	Max	06.05.1918	23.06.1918	POW
Uffz.	Stemmler	Gottfried	29.03.1918	11.04.1918	KIA
Lt.	Veith	Maximilian	17.06.1918	EOW	
Vzfw.	Wagner	Paul	06.11.1917	04.02.1918	KIC
Vzfw.	Walther	Siegfried	05.11.1917	27.04.1918	AFP 2 later Jasta 1F
Lt.d.Res.	Weig	Moritz	22.04.1918	02.05.1918	POW
Oblt.	Wenig	Erwin	06.11.1917	25.01.1918	Staffelführer Jasta 80

9. Airfields

Airfields	Arrival	Departure	Army
Habsheim near Mülhausen	04.11.17	25.03.18	Armee-Abteilung B
Liéramont	26.03.18	15.04.18	2. Armee
Suzanne sur Somme	15.04.18	09.07.18	2. Armee
Bignicourt	11.07.18	19.08.18	1. Armee
Sierenz (Sundgau)	19.08.18	05.09.18	Armee-Abteilung B
Habsheim near Mülhausen	18.09.18	end of war	Armee-Abteilung B

10. Jagdgruppen

Jagdgruppe*	Join	departure	Army	Jagdstaffeln
Jagdgruppe Süd	xx.03.18	???	2. Armee	Jasta 18, 58, 59, 76
Jagdgruppe C	11.07.18	19.08.18	1. Armee	Jasta 22, 63, 69, 76
Jagdgruppe	17.09.18 -	end of war	Armee-Abteilung B	Jasta 69, 71, 75, 76
* Non-permanent grouping of a number of Jagdstaffeln				

11 Identified Military Numbers of Aircraft Operated by Jagdstaffel 76

Aircraft	Pilot
Albatros D Va D.5294/17	Lt. Kolb
Albatros D Va D.5726/17	Lt. Emmerich
Albatros D Va D.5727/17 *	Lt. Hans Böhning / Vzfw. Karl Hopf
Albatros D Va D.5765/17	Lt. Bärenfänger / Lt. Schick
Albatros D Va D.5787/17	Vzfw. Josef Diem
Albatros D Va D.7221/17	Vzfw. Koller
Albatros D Va D.7249/17	Uffz. Stemmler
Albatros D Va D.7544/17	Uffz. Josef Bauer
Fokker D VII (O.A.W.) 6372/18	Oblt. Amandus Rostock
Fokker D VII (O.A.W.) 6381/18	Uffz. Geyer
Fokker D VII (O.A.W.) 6423/18	?, narrow white fuselage band
Fokker D VII (O.A.W.) 6481/18	Lt. Walter Curtius
Fokker D VII (Alb.) 5341/18	Lt. Veith
Pfalz D XII 1462/18	Uffz. Josef Bauer
Pfalz D XII 1516/18	Gefr. Julius Geyer

*Tentative identification of military number based on number deciphered on landing gear strut.

12 Excerpt from the War Diary of Royal Bavarian Jagdstaffel 76

07.09.17	Bavarian Flieger-Ersatz-Abteilung 1 in Schleißheim receives orders to complete formation of Jagstaffel 76 no later than October 15, 1917.
02.10.17	Lt. Walter Böning, Jagdstaffel 19, is ordered to Schleißheim and appointed Staffelführer of the soon-to-be established Jagdstaffel 76.
05.10.17	Lt. Walter Böning receives orders to proceed to Schleißheim.
07.10.17	Lt. Walter Böning arrives in Schleißheim.
01.11.17	The Staffel is ready to mobilize.
02.11.17	The Staffel departs Schleißheim at 19:30.
04.11.17	**The Staffel moves to Habsheim airfield near Mülhausen/Alsace in the German Army Division B Front Sector.**
05.11.17	Lt. Wilhelm Kolb and Vzfw. Siegfried Walther, both graduates of Jagdstaffelschule 2, are transferred to the Staffel.
06.11.17	Lt. Hans Böhning, previously Jagdstaffel 36, and Lt. Erwin Wenig, previously Jagdstaffel 28, are transferred to the Staffel.

15.11.17	Lt. Xaver Danhuber, previously Jagdstaffel 26, Vzfw. Karolus Bärenfänger and Vzfw. Paul Wagner, both previously Jagdstaffel 28, are transferred to the Staffel. Vzfw. Karolus Bärenfänger is awarded the Iron Cross 1st Class.
18.11.17	Lt. Clemens Vollert, previously Fliegerschule Lechfeld, is transferred to the Staffel as Offizier zur besonderen Verwendung (Special Task Force).
23.11.17	At 10:30, Vzfw. Siegfried Walther's departing Albatros D III (O.A.W.) collides with ground crew member Uffz. Eduard Feig. Feig is fatally injured by the propeller and dies during transport to Field Hospital 36. The pilot does not sustain any injuries.
01.12.17	Lt. Hans Böhning shoots down a Sopwith two-seater near Hagenbach. The aircraft makes an emergency landing in enemy territory. The downed aircraft is confirmed by Kogenluft his 5th aerial victory and the 1st Staffel victory, according to Bn. 114.591 Fl.III c from 13.12.17.
07.12.17	Vzfw. Siegfried Walther shoots down a twin-engined Caudron R XI above enemy territory near Gewenheim at approximately 15:45. The enemy aircraft goes down in flames. The downed aircraft is confirmed by Kogenluft as his 1st aerial victory and the 2nd Staffel victory.
	The downed aircraft was a Letord of the Escadrille F. 50; its crew, Brig. René Machetay (Pilot), S/Lnt. Maurice Mouillard (Observer) and Sold. René Michel (Machine Gunner), died.
09.12.17	Lt. Walter Böning shoots down a Spad single-seater between Thann and Masmünster at 16:17. The aircraft, engulfed in smoke, descends for a crash landing on enemy territory, near Giromagny. The downed aircraft is confirmed by Kogenluft as his 7th aerial victory and 3rd Staffel victory, according to Bn. 114.818 Fl. IIIc from 18.12.17.
12.12.17	Lt. Wilhelm Kolb fatally crashes in his Albatros D Va D.5294./17 south of Habsheim at 16:30, after the aircraft suffers wing failure. Kolb was on his return from an unsuccessful mission against the enemy. Uffz. Karl Hopf, Jagdstaffelschule 1, is transferred to the Staffel.
13.12.17	Lt. Xaver Danhuber is transferred as Flight Instructor to Fliegerschule 1 (flight school) of the Bavarian Flieger-Ersatz-Abteilung 1 in Schleißheim.
15.12.17	Uffz. Karl Hopf is awarded the Bavarian Militär-Verdienstkreuz III. Klasse mit Schwertern (Bavarian Military Cross of Merit, 3rd Class, with Swords).
17.12.17	Lt. Erwin Wenig receives word that he was selected to be given command of the soon-to-be formed Jagdstaffel 80. He is transferred to the Bavarian Flieger-Ersatz-Abteilung 2 in Fürth.
02.02.18	Lt. Walter Böning forces a Spad to land on enemy territory during aerial combat with 4 A.R.-two-seaters and their escort. Not confirmed as an aerial victory.
04.02.18	Vzfw. Paul Wagner fatally crashes at Colmar-north airfield at 10:46. Wagner crashes from low altitude, his aircraft catches fire upon impact, and burns completely. Lt. Walter Böning shoots down an enemy Spad; the aircraft crashes on enemy territory, between Füllern and Largitzen. The downed aircraft is confirmed by Kogenluft as his 9th aerial victory and 6th Staffel victory, in accordance with Bn. 117.418 Fl. IIIc from 23.02.18.
	The downed aircraft was a Sopwith of Escadrille Sop. 129; Pilot Sergt. Henri Cordonnier and Machine Gunner Cal. Allais died.
11.02.18	During aerial combat with two Paul Schmitt, Lt. Walter Böning shoots down one of the aircraft in flames at 16:50. It crashes near Füllern on enemy territory. Böning's aircraft receives a hit to the radiator, forcing him to make an emergency landing on German territory. The downed aircraft was confirmed by Kongenluft as his 8th aerial victory, and the Staffel's 5th victory, according to Bn. 117.614 Fl. IIIc from 22.02.18. The second Paul Schmitt is shot down by Uffz. Siegfried Walter in the same aerial battle; it crash lands near Carspach on enemy territory. Kogenluft confirms the downed aircraft as his 2nd aerial victory, and 5th Staffel victory.

	The downed aircraft were a pair of Dorand A.R. 2's of Escadrille A.R. 230. They were flown by M.d.L. Frédéric Monnier (Pilot) and S/Lnt. Taron (Observer), as well as S/ Lnt. Goeury (Pilot) and S/Lnt. Berger (Observer). As the first mentioned crew was injured (the observer sustained grave injuries), it is assumed that their aircraft was the one which was shot down in flames.
12.02.18	Uffz. Siegfried Walther is awarded the Iron Cross 1st Class.
13.02.18	Uffz. Josef Diem, Jagdstaffelschule 1, is transferred to the Staffel.
16.02.18	Lt. Hans Hertrich, Jagdstaffelschule 2, is transferred to the Staffel. Vzfw. Siegfried Walther is awarded the Ehrenbecher „Dem Sieger im Luftkampf" (Honorary Cup, "Victor in Aerial Combat").
18.02.18	Lt. Hans Böhning shoots down an A.R. 2 which crash lands near Dammerkirch on enemy territory. Kogenluft only acknowledges the downed aircraft as "forced to land beyond enemy front lines".
23.02.18	Lt. Hans Böhning is transferred to Jagdstaffel 79 and appointed Staffelführer of this unit. Gefr. Xaver Hüttner, Jagdstaffelschule 2, is transferred to the Staffel.
26.02.18	Uffz. Josef Diem hits an A.R. several times in aerial combat; the aircraft executes an emergency landing on enemy territory, descending "stinkend" (trailing smoke). Not confirmed as an aerial victory. Vzfw. Karolus Bärenfänger is promoted to Leutnant.
05.03.18	Gefr. Xaver Hüttner is promoted to Unteroffizier.
06.03.18	During aerial combat with 3 Spad single-seaters, Lt. Walter Böning shoots down two of the aircraft at approximately 13:45; both aircraft crash in enemy territory near Rodern. The first aircraft entered a spin, whereas the second aircraft caught fire prior to impacting the ground. The downed aircraft were confirmed by Kogenluft as 10th and 11th aerial victories, as well as the 7th and 8th Staffel victories, respectively, in accordance with Bn. 118.765 Fl. IIIc dated 22.03.18.
	The downed aircraft were two Spad XIII of the Escadrilles Spa. 150 and Spa. 151. The pilots, S/Lnt. Henri Dorizon (Spa. 150) and Cal. Paul Brun (Spa. 151), died.
08.03.18	Uffz. Siegfried Walther is promoted to Vizefeldwebel.
09.03.18	The Jagdstaffeln 41 and 76 are relocated.
10.03.18	The Jagdstaffeln 41 and 76 are moved back to Habsheim after the airfield was newly refurbished.
15.03.18	During aerial combat with three French aircraft, Lt. Walter Böning shoots down two Sopwiths in flames near Lautenbach, one of which crashes in French territory; the second aircraft crash lands between the front lines in no man's land. The downed aircraft were confirmed by Kogenluft as 12th and 13th aerial victories, as well as 9th and 10th Staffel victories, respectively, in accordance with Bn. 119.342 Fl. IIIc dated 01.04.18. During the same aerial battle, Uffz. Karl Hopf shoots down a Spad single-seater near Altthann above enemy territory. Kogenluft confirms the downed aircraft as his 1st aerial victory and the 11th Staffel victory.
	Both downed Sopwiths belonged to Escadrille Sop. 252. One of these, Sopwith 3476, flown by Sergt. Gacon (Pilot) and S/Lnt. Malot (Observer), crashed on German territory (both crew members died). The other Sopwith that made an emergency landing in no man's land was flown by Cal. Marcus (Pilot) and S/Lnt. Guérineau (Observer). Both crew members were able to save themselves by escaping back across their own lines.
18.03.18	Definitive transfer orders arrive at the Staffel.

19.03.18	Staffel transferred to Neuvilly, in the 2nd German Army's frontal sector. Upon arrival, the Staffel is relocated, finally setting up at Lièramont airfield, where, together with Jagdstaffeln 5 and 46, they form Jagdgruppe 2. This formation is put under the command of Oblt. Richard Flashar (Führer Jagdstaffel 5).
25.03.18	Lt. Richard Emmerich, Jagdstaffelschule 2, is transferred to the Staffel.
29.03.18	Uffz. Gottfried Stemmler, Jagdstaffelschule 2, is transferred to the Staffel.
31.03.18	Lt. Richard Emmerich is awarded the Iron Cross 1st Class.
03.04.18	Lt. Walter Böning shoots down a Sopwith Camel, which crashes approximately 300 m across the lines. The downed aircraft is confirmed by Kogenluft as his 14th aerial victory, and the 12th Staffel victory, in accordance with Bn. 121.391 Fl. IIIc dated 26.04.18. Lt. Walter Angermund, Jagdstaffelschule 1, and Vzfw. Karl Koller, Jagdstaffelschule 2, are transferred to the Staffel.
06.04.18	Lt. Hans Hertrich and Uffz. Xaver Hüttner are transferred to Armee-Flugpark 2. Lt. Walter Angermund is promoted to Oberleutnant.
07.04.18	Oblt. Walter Angermund and Vzfw. Karl Koller arrives at the Staffel.
11.04.18	Uffz. Josef Diem forces an R.E. 8 to land at 18:20 near Henancourt (?) across enemy lines. At 18:25, Lt. Walter Böning shoots down an S.E. 5a which catches fire, and carries out an emergency landing north-east of Albert in enemy territory. The downed aircraft are confirmed by Kogenluft as Lt. Böning's 15th aerial victory and the 13th Staffel victory, in accordance with Bn. 121.837 Fl. IIIc from 28.04.18. Uffz. Diem's opponent was recognized as "forced to land across enemy lines" only.
	The downed aircraft were R.E. 8 B844 from 15. Squadron RAF and S.E. 5a C5432 from 56. Squadron RAF. The crews of both the R.E. 8 (2/Lt. R G Hart (Pilot) and 2/Lt. L F Handford (Observer)) as well as the pilot of the S.E. 5a (2/Lt. H J Walkerdine) were injured.
	Lt. Richard Emmerich is shot down in aerial combat with an R.E. 8 at the controls of his Albatross D Va D.5726/17, breaking up in mid-air (abmontierend). The aircraft impacted the ground approximately 4 km west of Albert on enemy territory, where it burned completely. Uffz. Gottfried Stemmler, flying Albatros D Va D.7249/17, is shot down in flames around 18:25 in aerial combat against approximately 12 to 14 S.E. 5a aircraft. He fatally crashes about 4 km north-east of Albert on the German side of the lines.
	The opponents were the aforementioned crew from 15. Squadron RAF, shot down by Uffz. Diem, as well as probably Lt. T Durrant from 56. Squadron RAF (56. Squadron reported a total of three downed aircraft; the point of impact of his "Albatros scout" did not exactly match Uffz. Stemmler's crash site, but he was the only one to report an opponent "shot down in flames".)
14.04.18	Uffz. Ludwig Artmann, Jagdstaffelschule 2, is transferred to the Staffel.
15.04.18	The Staffel is relocated to Suzanne-sur-Somme airfield.
20.04.18	Vzfw. Siegfried Walther and Uffz. Karl Hopf are transferred to Armee-Flugpark 2.
22.04.18	Lt. Moritz Weig, Jagdstaffelschule 1, is transferred to the Staffel.
23.04.18	During an aerial battle between approximately 20 German and 20 English fighter aircraft, Lt. Walter Böning shoots down an S.E. in flames, which crashes north of Sailly-Laurette on its own territory. This victory claim, among others, was disputed between by Lt. Egon Koepsch, Jagdstaffel 4 (Jagdgeschwader 1); eventually, Kogenluft awarded credit for the aerial victory to Lt. Koepsch. Lt. Edmund Neu and Lt. Hermann Obermeier, both graduates of Jagdstaffelschule 2, are transferred to the Staffel.

29.04.18	Lt. Walter Böning takes a three-week leave; Oblt. Walter Angermund assumes deputy command of the Staffel in Böning's absence.
02.05.18	Lt. Moritz Weig loses contact to his Kette on a frontline flight, and fails to return. It is later reported that he was captured by the French. Uffz. Josef Diem is awarded the pilot´s badge (Flugzeugführer-Abzeichen).
04.05.18	Lt. Franz Martin and Vzfw. Georg Markert are transferred to the Staffel, coming from Jagdstaffelschule 1.
06.05.18	Lt. Max Schick, Jagdstaffelschule 1, is transferred to the Staffel.
11.05.18	Lt. Hermann Obermeier is injured following an emergency landing; he is transported to field hospital 90. The reason for the emergency landing is not recorded in the available documents.
17.05.18	Vzfw. Karl Koller shoots down a D.H. 4 in flames north-west of Mesnil in enemy territory. The downed aircraft is confirmed by Kogenluft as his 1st aerial victory, and the Staffel's 14th victory.
20.05.18	Lt. Walter Böning returns from his leave and resumes his duties as commander of the Staffel.
21.05.18	Lt. Walter Böning is awarded the Knights Cross of the Hohenzollern House Order (Ritterkreuz des königlichen Hausordens von Hohenzollern).
23.05.18	Oblt. Walter Angermund is injured in aerial combat with English fighter aircraft; he carries out an emergency landing near Fricourt. As a result, his aircraft is destroyed, causing Angermund further injuries: he is transported to a field hospital.
25.05.18	At 20:00, Vzfw. Karl Koller departs on a fighter patrol flight with his Kette in Albatros D Va D.7221/17. He does not return from the flight. Following aerial combat with English fighters, Koller was forced to make an emergency landing on enemy territory, and was subsequently taken captive.
	Hi opponent was Lt. Alfred Alexander Leitch of 65. Squadron RAF, flying Camel D1903; Leitch reported an "Albatros scout" shot down near Senlis *on the Allied side at 19:50 (British time).*
28.05.18	Uffz. Josef Diem forces an S.E. 5a to land on enemy territory following aerial combat near Hangard. Kogenluft only recognizes the "aircraft forced to landing beyond enemy lines". Lt. Franz Martin sustains light injuries during the same aerial battle; he was able to return to his airfield and land safely.
	The opponents were likely S.E. 5a´s from 84. Squadron RAF who reported four "Albatros scouts" shot down (1 "crashed" and 3 "out of control") between 13:25 – 13:40 (British time). Lt. Franz Martin's Albatros was likely one of the aircraft reported to go down "out of control". Lt. B. Stefanson was injured in the same battle, flying S.E. 5a D333. He could have been Uffz. Diem's opponent: this information is unconfirmed, however, as the exact time at which Lt. Stefanson was injured has not been recorded.
29.05.18	Uffz. Ludwig Artmann is awarded the pilot's badge (Flugzeugführer-Abzeichen).
30.05.18	During a Staffel sortie around midday, Lt. Walter Böning shoots down a Camel which makes an emergency landing on its own territory near Contalmaison. In the same aerial battle, Vzfw. Georg Markert pursues an additional Camel aircraft across the lines, and shoots it down in flames at approximately 12:30 near Albert on enemy territory. The Kogenluft confirmed the downed aircraft as their 16th aerial victory and 15th Staffel victory, as well as the 1st aerial victory and 17th Staffel victory, respectively, according to Bn. 125.504 Fl. A from 17.06.18.

	The opponent was a Flight of the 70. Squadron RAF who reported two "Albatros scouts" shot down in aerial combat between 11:10 – 11:20 British time (1 "crashed", and 1 "out of control"), east of Albert. 70. Squadron suffers the loss of Lt. W A Scott in Camel D6640, who makes an emergency landing on German territory, where he is taken into captivity.
	While on an afternoon Staffel mission, Lt. Walter Böning shoots down a Camel in flames near Bray on our own territory, the aircraft crashes near its own airfield. The downed aircraft is confirmed by the Kogenluft as his 17th aerial victory and the 16th Staffel victory, according to Bn. 125.504 Fl. A dated 17.06.18.
	The opponent was a Flight from 43. Squadron RAF who reported no less than seven "Albatros scouts" shot down in aerial combat between 14:50 – 15:00, British time (4 "crashed", and 3 "out of control"), near Flers-Combles. 43. Squadron loses Lt. P T Bruce in Camel D1793, who is shot down and killed over German territory.
31.05.18	Uffz. Josef Diem shoots down a Bréguet Br. 14 in flames over enemy territory near Hangard. The downed aircraft is confirmed by the Kogenluft as his 1st aerial victory, and the 18th Staffel victory.
	The Staffel is attacked by a superior "S.E.-Formation" while on their last sortie of the day, shortly after departure while still climbing: Lt. Walter Böning, flying Albatros D Va D.5765/17, is badly wounded by a stray bullet; he manages to return to his airfield and lands safely. Moments after Vzfw. Georg Markert dug into one of his opponents, he was attacked by another from behind, who shoots him down in flames at approximately 18:15. He crashes near Fricourt, is taken from the wreckage of his aircraft but dies the next day.
	The opponent was probably a Flight from 70. Squadron, RAF. Flight Commander Capt. J Todd in Camel C1670 reported an "Albatros Scout" shot down in flames south-east of Albert at approximately 18:00 (British time). Walter Böning's "departure" did not appear to be noticed by the British during the confusion of the encounter.
01.06.18	*Lt. Max Schick takes over Command of the Staffel as deputy.*
07.06.18	Lt. Max Schick shoots down a "British Spad Single-Seater". The aircraft crashes near Millencourt in enemy territory. Kogenluft confirms the downed aircraft as his 1st aerial victory, and the 19th Staffel victory.
17.06.18	Arrival of Lt. Maximilian Veith, who is transferred from Jagdstaffelschule 1 to the Staffel.
21.06.18	Lt. Karolus Bärenfänger is transferred to Flieger-Ersatz-Abteilung 1 in Schleißheim. Arrival of Lt. Karl Dumler who is transferred from Jagdstaffelschule 1 to the Staffel.
22.06.18	Uffz. Josef Diem is awarded the Iron Cross.
23.06.18	The members of the Staffel learn that Lt. Ludwig Schmid, Jagdstaffel 34, has been named the new Staffelführer. Lt. Max Schick departs on a fighter patrol flight at 20:15 in Albatros D Va D.5767/17 from which he fails to return. According to reports received later, he was taken into captivity.
	According to the French Aéronautique Militaire war diaries, the IIIéme Armée's anti-aircraft artillery shot down a German single-seater on the Picardie front sector on that day, whose pilot was taken prisoner of war.
24.06.18	Lt. Karl Neu and Lt. Max Schick are awarded the pilot's badge (Flugzeugführer-Abzeichen).
05.07.18	Gefr. Johann Brunner and Josef Reinhard, both graduates of Jagdstaffelschule 1, are transferred to the Staffel.

08.07.18	Lt. Walter Curtius, Jagdstaffelschule 1, is transferred to the Staffel.
09.07.18	Staffel moves to Bignicourt in the 1st German Army's Frontal Sector. There, they form Jagdgruppe C, consisting also of Jagdstaffeln 22, 63, and 69. Jagdgruppe C is put under the Command of Lt. Alfred Lenz (Führer Jagdstaffel 22). Lt. Max Schick is awarded the Ehrenbecher „Dem Sieger im Luftkampf" (honorary cup, "Victor in aerial combat").
12.07.18	At approximately 21:00, Uffz. Ludwig Artmann fatally crashes about 2 km north of Bignicourt airfield, shortly after takeoff, from an altitude of 400 m as a result of wing failure.
16.07.18	Lt. Ludwig Schmid is badly injured during an attack on a captive balloon by defending ground fire. He manages to return back over the lines, where he makes an emergency landing. Lt. Schmid is transported to a field hospital. Lt. Karl Neu is injured in aerial combat against several Spad single-seaters. Following an emergency landing in friendly territory he is taken to a field hospital.
22.07.18	Following recovery, Lt. Hermann Obermeier returns to the Staffel.
23.07.18	Uffz. Josef Bauer und Gefr. Julius Geyer both arrive at the unit, coming from Jagdstaffelschule 2.
27.07.18	Lt. Hans Hertrich, Bavarian Flieger-Ersatz-Abteilung 1, is transferred to the Staffel.
29.07.18	Uffz. Rupert Hainzl, Jagdstaffelschule 2, is transferred to the Staffel.
30.07.18	Uffz. Edwin Hepperlin, Jagdstaffelschule 1, is transferred to the Staffel.
05.08.18	Lt. Franz Martin receives orders to join the Aflup. F in Palestine and leaves the Staffel.
08.08.18	Lt. Eberhard Grosse, Jagdstaffel 72, is ordered to join the Staffel as deputy Staffelführer.
12.08.18	Following recovery, Lt. Karl Neu returns to the Staffel.
18.08.18	Uffz. Josef Bauer is shot down after aerial combat against 12 to 14 enemy fighters near Château Thierry, flying Albatros Va D.7544/17. He is able to bring down the aircraft on friendly territory with a dead engine. The aircraft is completely destroyed, the pilot escapes unharmed.
19.08.18	Staffel moves to Sierenz in the Frontal Sector of the German Army, Division B. The Staffel is equipped with Pfalz D XII.
01.09.18	Lt. Hans Hertrich is put at the disposal of the Bavarian Aviation Inspection.
17.09.18	Orders are issued to combine Jagdstaffeln 69, 71, 75, and 76 into Jagdgruppe Habsheim. Jagdgruppe Habsheim is under the command of Oblt. Alexander Thomas, Führer of Jagdstaffel 69.
18.09.18	Staffel moves back to Habsheim. Lt. Hermann Obermeier, Gefr. Johann Brunner and Gefr. Josef Reinhard are awarded the "pilot's badge" (Flugzeugführer-Abzeichen).
06.10.18	Staffelführer Lt. Ludwig Schmid, who is still in the process of recovering from his injuries and therefore absent from the Staffel, is officially transferred to the Bavarian Flieger-Ersatz-Abteilung 1 in Schleißheim.
07.10.18	Oblt. Amandus Rostock, Jagdstaffel 77, arrives to the Staffel as its new Staffelführer. Lt. Eberhard Grosse's temporary duties are repealed, and he returns to Jagdstaffel 72.
10.10.18	Lt. Karl Dumler and Lt. Maximilian Veith are awarded the "pilot's badge" (Flugzeugführer-Abzeichen).
11.10.18	Gefr. Julius Geyer is promoted to Unteroffizier.

15.10.18	Gefr. Johann Brunner and Josef Reinhard are promoted to Unteroffizier.
30.10.18	Lt. Karl Dumler shoots down a D.H. 9 who crash lands west of Carspach in enemy territory. The downed aircraft is confirmed by Kogenluft as his 1st aerial victory.
	The downed aircraft was a Salmson 2A2 of Escadrille Sal. 41. Its crew, Sergt. Loriat (Pilot) and Soldat Monjalat (Machine gunner) were both injured; the gunner sustained severe injuries.
03.11.18	Lt. Walter Böning arrives at Habsheim to take over the leadership of Jagdstaffel 76, but since no corresponding order from Munich has arrived at Habsheim at this point, Lieutenant Rostock continues to lead the squadron.
06.11.18	Lt. Walter Böning leaves Habsheim and returns to Munich
11.11.18	Armistice
14.11.18	The members of the Jasta 76 leave Habsheim for Munich. The aircraft are left behind at Habsheim airfield.

13. The Research

In 1974 I obtained a small booklet listing the names and the addresses of the members of the "Alte Adler" (Old Eagles, the association of German pre- and First World War aviators). I sorted out the members of the Jagstaffeln and then tried to get in contact with the old gentlemen. In the late spring 1976 I had my first phone call to Dr. Walter Böning, former member of Feldflieger-Abteilung 6b, Jagdstaffel 19 and Staffelführer of Jagdstaffel 76. Like most of the old pilots I got in touch with he was very much surprised that, after the passage of roughly sixty years, suddenly someone showed an interest in his career as a pilot in World War I. He invited me to visit him, and in September 1976 I seized the chance to meet him at his home in the city of Arolsen.

Dr. Walter Böning, now 82 years of age, was a very active person who had worked as a physician until the end of his 70´s. He was mentally fully active, and only ocassionally suffered from the late effects of the wound in his leg from the 1st World War. I was very warmly welcomed by him and his wife, as well as their son in law Dr. Figge, who was also his successor as head of the local hospital. I stayed for two days and Dr. Böning made his three photo albums available to me for research. I used the time to go through the photo albums with Dr. Böning in order to obtain additional information from him about the photos. A point of particular interest for me was, of course, the painting schemes of the aircraft shown in the photos.

He also showed me the letters he had written to his parents during the war, from 1914 – 1918. He told me it was the best way for him to deal with the daily events in the war, when on the evening he could write his experience down. Writing these letters brought a kind of relieve to him. Luckily, his parents retained all the letters he had written. Some years after the war they asked someone to typewrite all the letters and got them bound as a book as a Christmas present to him. Dr. Figge very kindly promised me to copy the book, which had 370 pages.

Two week later I received the copy of the letters and I started to read them. With the letter in my hands, I recognised the historical value of these letters, as they documented substantial parts the history of Bavarian Feld-Flieger-Abteilung 6, Jagdstaffel 19 and Jagdstaffel 76. They contained countless personal experiences, as well as information never to be found in any official document. With a great deal of astonishment, I read how much difficulty a Staffelführer had to face when he had to organize the formation of a Jagdstaffel under the so-called America-program in Bavaria.

As I found the material extremely interesting, I made a second trip to Arolsen in October 1976 to meet Dr. Walter Böning again. In this case I stayed nearly three days and he took the time to meet me every day. The focus of the second meeting was to go with him to go through his letters, dealing mainly with Jagdstaffel 19 and Jagdstaffel 76, in an attempt to collect additional information. For me it was obvious that this extraordinary material, and the information contained therein had to be published one day. Dr. Walter Böning was a fascinating person and possessed the qualities of a natural a born leader. It was fascinating to see that, even after nearly six decades, there was still a fire in his eyes when he spoke about his time as a fighter pilot and Staffelführer. This created a special feeling which took me back to the days of his World War I aviation adventures. Without the massive amount of highly detailed information he provided, and his generous and friendly support, this book would have never been written!

Some years later I was able to establish contact to Frau Anna Vollert, the widow of Herrn Clemens Vollert, the special duties officer of Jagdstaffel 76. The very friendly and helpful lady also gave me the chance to copy the photo albums of her late husband, which contained photos of Bavarian Feld-Flieger-Abteilung 6 and Jagdstaffel 76. Surprisingly, she also had a photo-album with very interesting photos covering the early days of Jagdstaffel 2, and I could also copy these pictures.

After time-consuming research, I also found the daughter of Jagdstaffel 76 pilots Josef Bauer, Mrs. Luise Wildmann as well as relatives of Julius Geyer and Karl Hopf, who also kindly provided access to their photos and notes. Especially the photos and information left behind by Julius Geyer and Josef Bauer provided interesting details about the the aircraft equipment of Jagdstaffel 76 during the closing stages of the war. Finally, Herr Polizeipräsident a.D. Koller was so kind to give me the chance to copy the photos of his late father.

All of this research provided a very good insight into the events from the German perspective but collecting information on the opponents of Jagdstaffel 76 was also a challenge. In a good number of cases, the adversaries of Jagdstaffel 76 in aerial combat could be identified, but of course that was not possible in every case. Identifying an enemy in aerial combat is a very difficult and puzzling matter, and sometimes only a likely opponent can be singled out. On the French front, this is made even more difficult by the fact that French records are incomplete due to the destruction of some archival material during the 2nd World War. The book "The French Air Service War Chronology", published in 2001, serves as a good aid in this quest, but unfortunately does not include all information available from French sources.[1] This is shown by a comparison with the contemporary French publications in L'Aérophile and other sources, e.g., La Guerre Aérienne illustrée, websites such as "Mémoire des Hommes", reports in French magazines and newspapers, local and regional casualty lists of French cities and regions and various private websites on French airmen. Nevertheless, for all those interested in the French Aviation Service in World War I, this book is a valuable, if incomplete, source.

The book was nearly finished when photos of Jagdstaffel 76 appeared for sale on an online auction website. Fortunately, Greg VanWyngarden, Reinhard Kastner and Rainer Absmeier were able to obtain these photos and thereby rescued them for historical research, rather them letting them disappear for ever in the hands of an unknown collector. They very generously made available their precious photographs for this book. These photos, as well as photos of surrendered aircraft of Jagdstaffel 76 taken at Trier and Romorantin in late 1918 and 1919, provided by Greg VanWyngarden, Jörn Leckscheid and Tobias Weber, allowed for a comprehensive overview of the aircraft operated by Jagdstaffel 76 during the last months of the war.

The war diary of Jagdstaffel 76 that was stored in the Reichsarchiv in Potsdam in the inter-war years ended in June 1918. It is very likely that the last volume of the war diary was taken home by a member of the unit following the armistice, perhaps by the last Staffelführer, and never delivered to the Reichsarchiv, as this also the case with the war dairies of some other units. Winfried Bock and Reinhard Kastner reconstructed the events of the missing timeframe by recovering information from several available documents and information held by the Bavarian Main State Archive, a work that spanned several years. But of course, we do not know how complete this information is.

It is in this sense that I hope that the book presents the short history of Jagdstaffel 76 in the most comprehensive manner possible. Hopefully, it will also preserve the memories of the young pilots who flew, fought and died in their aircraft made of fabric, steel and wood.

Bruno Schmäling, Autumn 2021

Right: Bruno Schmäling together with Dr. Walter Böning and his wife in front of their home in Arolsen in September 1976.

28.8.31

21. 11. 17. (23.11.) ... Vorläufig regnet es hier ziemlich dauerhaft. Im Tal zwischen 3 Gebirgen ist es nur bei Nordwinden "fliegerisch". Im Grunde genommen wars mir für die erste Zeit ganz recht, denn bis die Maschinen eingeschossen und ganz hergerichtet sind, von z.T. ganz unerfahrenen Leuten, vergeht doch einige Zeit. Wirkliches Glück habe ich, soweit ich es bisher beurteilen kann, mit meinen Flugzeugführern gehabt, die fliegerisch meist glänzend sind und kameradschaftlich so, wie man es sich nicht besser wünschen kann. Eigentümlicherweise spielt ausser mir fast keiner Karten, sodass nicht mal ein Tarok zusammen geht; dafür sind aber alle derart musikalisch, dass ich fast neidisch werden möchte. Zwei Herren spielen Guitarre, 2 andere Mandoline, einer Klavier und einer Geige. Ausserdem ist noch eine Handharmonika in Betrieb und mehrere sind Dilettanten am Klavier. Abends geht es da bei uns ganz lustig zu, und ich bin froh, dass sich die Lustigkeit in diese Bahnen lenkt.

Ich habe mir verschiedene Herren "persönlich" angefordert, einen Adjutanten und einen techn. Offizier (Lt. Böhning aus Bremen), mit denen ich schon lange gut befreundet bin, und habe sie bekommen. Es wird Euch jetzt bald so scheinen, als ob da nichts zu wünschen übrig bliebe, aber es gibt doch noch sehr viele Dinge, die nicht nach Wunsch sind. Vor allem habe ich augenblicklich einen stellv. Kofl., der Pedant und Kleinigkeitskrämer bis zum Aeussersten ist. Es ist natürlich, dass manche schriftl. Sachen vorerst noch kleine Formfehler haben etc.. Die Folge davon ist, dass Schreiben, die sehr beschleunigt werden müssten, nochmals zurückkommen und nochmal geschrieben werden müssen, weil ein überflüssiges Wort darin stand und "damit die Papiernot behoben wird". Ein ständiger Aerger bleibt auch der Materialnachschub, entweder bekommt man nur einen kleinen Teil von dem, was man "dringend" braucht, oder man bekommt "Ersatz" oder nichts taugl.

306

Typewritten transcript of Walter Böning's diary.

28.8.31

21. 11. 17. (23.11.) ... Vorläufig regnet es hier ziemlich dauerhaft. Im Tal zwischen 3 Gebirgen ist es nur bei Nordwinden "fliegerisch". Im Grunde genommen wars mir für die erste Zeit ganz recht, denn bis die Maschinen eingeschossen und ganz hergerichtet sind, von z.T. ganz unerfahrenen Leuten, vergeht doch einige Zeit. Wirkliches Glück habe ich, soweit ich es bisher beurteilen kann, mit meinen Flugzeugführern gehabt, die fliegerisch meist glänzend sind und kameradschaftlich so, wie man es sich nicht besser wünschen kann. Eigentümlicherweise spielt ausser mir fast keiner Karten, sodass nicht mal ein Tarok zusammen geht; dafür sind aber alle derart musikalisch, dass ich fast neidisch werden möchte. Zwei Herren spielen Guitarre, 2 andere Mandoline, einer Klavier und einer Geige. Ausserdem ist noch eine Handharmonika in Betrieb und mehrere sind Dilettanten am Klavier. Abends geht es da bei uns ganz lustig zu, und ich bin froh, dass sich die Lustigkeit in diese Bahnen lenkt.

Ich habe mir verschiedene Herren "persönlich" angefordert, einen Adjutanten und einen techn. Offizier (Lt. Böhning aus Bremen), mit denen ich schon lange gut befreundet bin, und - habe sie bekommen. Es wird Euch jetzt bald so scheinen, als ob da nichts zu wünschen übrig bliebe, aber es gibt doch noch sehr viele Dinge, die nicht nach Wunsch sind. Vor allem habe ich augenblicklich einen stellv. Kofl., der Pedant und Kleinigkeitskrämer bis zum Aeussersten ist. Es ist natürlich, dass manche schriftl. Sachen vorerst noch kleine Formfehler haben etc. Die Folge davon ist, dass Schreiben, die sehr beschleunigt werden müssten, nochmals zurückkommen und nochmal geschrieben werden müssen, weil ein überflüssiges Wort darin stand und "damit die Papiernot behoben wird". Ein ständiger Aerger bleibt auch der Materialnachschub, entweder bekommt man nur einen kleinen Teil von dem, was man "dringend" braucht, oder man bekommt "Ersatz" oder nichts taugl.

306

Typewritten transcript of Walter Böning's diary.

14 The Further Lives of the Pilots of Jagdstaffel 76

Walter Böning

Right after the end of the war, Böning was commanded to Schleißheim in order to serve there in the capacity of a demobilization officer, and later with the equipment company. After being sent on leave in January, he was promoted Oberleutnant of the reserves on the last day of the month. On April 22, 1919, he volunteered to join the Bavarian Rifle Corps, which was later re-designated Bavarian Rifle Brigade 21. The unit was commanded by Oberst Franz Ritter von Epp, and participated in the advance of the Freikorps against the Soviet Republic in Munich.

He resumed his medical studies after completion of his military service, and later settled down in Arolsen to work as a doctor. There he became a senior physician and later took up the position of head of the hospital. Even after his retirement, he continued to work as a practicing physician until the end of his seventieth year of life, taking care of his elderly patients. On February 7, 1981, Dr. Walter Böning passed away in Arolsen.

Walter Angermund

After the end of hostilities, Oberleutnant Angermund was commanded to the Luftfahrt-Friedens-Kommission (Aviation Peace Commission). Since he prevented the localization of airframe and engines, he was discharged from the Reichswehr at the instigation of the international escort officers. From 1922 to 1924 he was employed by Junkers Air Traffic, Angermund was a senior official of the Trans-Europa-Union there and left the airline at the end of 1924 and joined Udet-Flugzeugbau, based in Munich-Ramersdorf. A year later he was employed by Lufthansa, working as the head of the company´s press office. For his invention of a device handling simultaneous loading and unloading of freight he was awarded a patent in 1929.

After 1933, he became the propaganda director of the „Deutschen Luftsport-Verbandes" (German Air Sports Association) and editor of the magazine "Die Luftwelt" ("The World Of Aviation"). After being re-called into active service in 1937, he joined the Reichsluftfahrt-Ministerium (RLM - Reich Aviation Ministry). Here, he became the personal assistant to the Generalluftfahrzeugmeister (general aircraft master) Ernst Udet. During the course of the war, he was promoted Oberst of the Luftwaffe

Following the end of the war, Angermund was taken into captivity by the Allies. Following his discharge, he lived in Vilsbiburg, where he passed away in October 1976, aged 83.

Karolus Bärenfänger

Following his transfer to the Flieger-Ersatz-Abteilung 1 in Schleißheim, he was first assigned to the Military Flying School 1 as a technical officer and was then transferred to the Military Flying School 4 in Lager Lechfeld in July 1918. Here he was also able to pass on his experiences as a fighter pilot to the pilots undergoing training training. Shortly before the end of the war, he was assigned to Jagdkette Schonger on 3 November 1918.

On March 2, 1919 he retired from the army and moved to Halle an der Saale. From June 16, 1919 he worked there for the Adlerwerke, first as a representative, later as a general agent for various branches of the Adlermobil sales company in Bremen and Munich. In 1931 he left the company and worked until 1933 as a freelance representative for various foreign companies. In January 1933 he was employed as a representative at the Munich general agency of the Magiruswerke. In the same year he became a member of the German Air Sports Association in Munich. Like many of his former comrades, he enlisted in the new German Luftwaffe after its founding in 1935. He served in various Luftwaffe units as a commander. On August 1, 1944 he was promoted to Oberst (colonel) and was taken prisoner at the end of the war. Further information on the further fate is not known to the authors. His son Karl died as a lieutenant and pilot at 7./KG 27 on July 29, 1942 along with the crew of his airplane.

Josef Bauer

After demobilization in Sonthofen, he was transferred to Fürth on January 21, 1919 to the Bavarian Flieger-Ersatz-Abteilung 2 and was eventually discharged from military service on May 11, 1919. He worked as a professional jockey, horse trainer and riding instructor. He remained a citizen of Munich until the end of his life.

Hans Böhning

As leader of Jagdstaffel 79 he scored another 12 (amounting to a total of 17) aerial victories up until September 20, 1918. The last of these was not officially confirmed until the end of the war. On September 20, 1918, Jagdstaffel 79 took off from its airfield at Guise-Ost in order to intercept incoming British De Havilland D H 9 bombers. In the ensuing dogfight, he was seriously wounded, but was still able to land his plane safely back at the airfield of his Staffel.

After his recovery, he was appointed squadron leader of the Royal Bavarian Jagdstaffel 32 on November 1, 1918. However, at this late stage of the war, he could only undertake a few front-line flights with his squadron. After the armistice came into effect he, and several fellow pilots of his Jasta tried to fly their planes to the Flieger-Ersatz-Abteilung at Fürth. Along with a few of them, he succeeded in this endeavour.

He then began studying mechanical engineering and engineering at the Technical University at Darmstadt. He married his fiancée Agnes Janson. The couple had three sons in the following years. Returning to Munich, together with the fighter pilot Eduard Ritter von Schleich and the engineer Werner von Langsdorff, he was one of the founding members of the "Leichtflugzeug Club München" which designed and built gliders.

He successfully participated in various European flight competitions, including the "International Trans-European Flight" in 1930 in the cockpit of his B.F.W (Bayerische Flugzeug Werke) M (Messerschmidt) 23b. On October 20, 1934, he fell victim to an accident during the maiden flight of the glider "Zögling" on an airfield at Pasing near Munich, and he succumbed to his injuries two days later in the hospital in Pasing. More than 1.000 people attended the funeral and followed the coffin of the popular aviator.

Johann Brunner

After the end of the war and the disbanding of the Staffel in Sonthofen, he was discharged from military service on January 25, 1919. No details concerning his later life are available to the authors.

Walter Curtius

After the end of the war his Staffel was disbanded in Sonthofen, and he was discharged from military service on January 19, 1919. Nothing is known to the authors about his further life.

Franz Xaver Danhuber

After he left Jagdstaffel 76 and was transferred to Flieger-Ersatz-Abteilung 1, he was entrusted with the formation of the Bavarian Jagdstaffel 79. On January 15, 1918 he was appointed leader of the squadron, and the unit departed Schleißheim on January 30, 1918 and moved into the field. Already on February 11, 1918 Danhuber had to be admitted to the hospital following a crash at Thugny airfield, no enemy action was the cause of this accident. On June 11, 1918 he was assigned to the Bavarian Military Aviation School 1 in Schleißheim, and on October 2, 1918 he was again appointed Staffelführer of Jasta 79, which he then led until the end of the war. Because of his many war injuries, he was in need of outpatient treatment from January 5, 1919 to April 14, 1919. On May 19, 1919 he was discharged from military service but was reactivated for military service on August 4, 1919.

In Schleißheim he was assigned to the Bavarian Artillery Fliegerstaffel 121. On September 14, 1919, he finally ended his military career. In the post-war period, his centre of life changed and he moved to Hamburg and practiced his original profession as a teacher again. He died on October 4, 1960.

Josef Diem

Unfortunately, very little information about Josef Diem is available. On November 26, 1918, after the Staffel had been disbanded, he was transferred to the replacement company of the Bavarian Flieger-Ersatz-Abteilung 1 in Schleißheim and released into civilian life there. In 1940 he lived near Schongau in Upper Bavaria. Further details about Diem are not available to the authors.

Karl Dumler

After the end of the 1st World War, he was transferred to Bavarian Flieger-Ersatz-Abteilung 2. Karl Dumler retired from military service on November 27, 1918 and returned to Augsburg to join his parents' business. He enlisted in the new German Luftwaffe, soon after its creation and was deployed in various positions in the 2nd World War. On April 1, 1943, he was promoted to Major of the Luftwaffe Reserve. Nothing is known to the authors about his further life.

Julius Geyer

After demobilization of the Jagdstaffel in Sonthofen, he was awarded the Bavarian Pilot´s Badge on December 8, 1918 and the Military Service Badge III. Class on January 10, 1919. On June 20 ,1919 he was discharged from military service and returned to his hometown Donauwörth. There he joined his parents' construction company and took over the company after his father retired from the business.

Eberhard Grosse

Unfortunately, little information concerning the later life of Grosse is available. All that is known is that he was reactivated in the 1930s and served as a flight staff engineer in the new Luftwaffe. Nothing is known to the authors about his further life.

Edmund Neu

After the Staffel was disbanded, he was transferred to the Sonthofen airfield commandant's office. On January 7, 1919, he was transferred to Bavarian Infantry Regiment 23 and retired from military service. He lived in Ludwigshafen am Rhein in 1935 and was reactivated to the Luftwaffe in the 1930s. In November 1941 we find him in Piacenza in Italy. He was promoted to the rank of major of the reserves on June 1, 1943. Further details about him are not available to the authors.

Rupert Hainzl

After demobilization in Sonthofen, he was transferred there on November 26, 1918 and transferred to Bavarian Flieger-Ersatz-Abteilung 2 on January 21, 1919 and discharged from military service. After the 1st World War, he lived in Berlin W, Luitpoldstraße 37. He is said to have emigrated to the USA in the inter-war period.

Edwin Hepperlin

After the end of the war, he was awarded the Bavarian Pilot Badge on December 8, 1918 and retired from military service on March 25, 1919. Apparently, he had been a staunch supporter of National Socialism, because after the 2nd World War charges were brought up against him related to his Nazi past. Further information concerning him is not available to the authors.

Hans Hertrich

After returning to the Bavarian Flieger-Ersatz-Abteilung 1, he was transferred to the Military Flying School 5 in Gersthofen, where he saw the end of the war. He was discharged from military service on January 5, 1919 and returned to Nuremberg where he resumed his profession as a teacher. Reactivated in the 2nd World War, he served as a Hauptmann (captain) in the Luftwaffe and returned to his hometown Nuremberg after the end of the war. He passed away in Nuremberg on February 24, 1975.

Karl Hopf

Following his transfer to Armee-Flugpark 2 and then on to Bavarian Flieger-Ersatz-Abteilung 1 in Schleißheim, he moved on to the Aviation Gunnery School at Lechfeld. On July 27, 1918 followed the transfer to Armee-Flugpark 19 and from there he transferred to Bavarian Jagdstaffel 80, which was led by Lieutenant Erwin Wenig. After the end of the war and demobilization of the Staffel in Fürth, Hopf was on November 28, 1918 transferred to the Ersatz-Kompanie of the Flieger-Ersatz-Abteilung 2 and on December 13, 1918, he was discharged from military service. He returned to his hometown of Würzburg after the war.

Xaver Hüttner

Unterofffizier Hüttner was transferred to Armee-Flugpark 2 on April 6, 1918, followed by the transfer to the homeland to the Aviation Gunnery School in Lechfeld. On the July 27, 1918 he was assigned to Bavarian Jagdstaffel 78 but already on the August 24, 1918 he was back at Armee-Flugpark A. He left the Fliegertruppe and was transferred to the 2nd Company of the Infantry Leib-Regiment on September 25, 1918. In the turmoil of the Munich Council Republic, he registered to join a volunteer company that disbanded after the end of the revolution and evolved into the Bavarian Infantry Regiment 45. On July 13, 1919 he was released from military service. The authors lack information concerning his further life path.

Karl Koller

Karl Koller returned from British captivity on October 22, 1919. He left military service in December 1919 and joined the Bavarian state police. There he was deployed at the Bavarian Police Aviation Squadron 1. He also remained loyal to aviation outside his profession and devoted himself to gliding in the "Bavarian Aero Club". In 1921, he represented the state of Bavaria at the gliding competition on the Rhön with an experimental construction of his fellow citizen of Glonn, Emeran Stadler. Their design flewd without the usual vertical, side and aileron: the control was only carried out by wing-warping. Being able to cover a distance of 640 meters on August 15, 1921, this was the longest flight distance reached by a glider to date. Already on August 20, 1921, however, he set a new milestone with the new world record of 1900 meters between take-off and landing site, the flight duration was a much-noticed three minutes. Five days later, he was able to put up another performance: the first complete circle (and another curve of 180°) without loss of altitude with a glider. These flights earned him worldwide attention, he received teaching and demonstration assignments from England, France, Switzerland and other countries.

On August 1, 1935 he joined the Luftwaffe as Hauptmann, exactly one year later followed the promotion to the rank of Major. From February 1, 1939 he held the position of the head of the General

Staff of Luftflotte (Air Fleet) 3. He rose quickly through the military ranks and on November 1, 1944, he was appointed head of the General Staff of the Luftwaffe. Following the end of the war, he was imprisoned, from which he was released in December 1947. After his discharge, Koller returned to Glonn and began a professional restart as a representative of Volkswagen AG in Munich. In 1949, he published the personal memories of his experiences in the closing month of the 2nd World War under the title "The Last Month". On December 22, 1951 Koller died of a heart disease in Glonn at the age of 53.

Franz Martin

Shortly before the Armistice, he returned to the Staffel, and greeted the end of the war at the front. After his return home, he left the military on November 27, 1918.

Following the formation of the Luftwaffe, he was re-activated as an Officer, and was promoted Major of the Reserves on April 1, 1943. By August 1944 he had been assigned to the Leadership at E(v) 231/XII airfield in Bordeaux-Merignac.

Josef Reinhard

After the demobilization of the Staffel at the Kommandantur Sonthofen airfield, he was transferred to Flieger-Ersatz-Abteilung 2 in Fürth on January 14, 1919, there he was assigned to Military Aviation School 3. In the turmoil of the riots in and around Munich, he reported to the voluntary aviation department Krug on May 1, 1919. The unit was renamed Truppen-Fliegerstaffel (Troop Aviation Squadron) 21a on 1 June 1919.

At the end of September 1919, he was absorbed into the Reichswehr and assigned to the Schleißheim airfield. After the disbanding of all aviation units, he was first assigned to the News Department 24 (Horchabteilung 324) in April 1920 and in September of the same year to Department 7 (Driver Company 124). As of December 31, 1920 he left military service. Further life data is not known to the authors.

Amandus Rostock

After the formation of the new German Luftwaffe he re-entered service in 1935. During the 2nd World War he held the rank of Major of the Reserves, and from March 1943 he served with the Staff of Kampfgeschwader I. In October 1943 he joined the 2nd Fliegerdivision as Ic. After the 2nd World War he lived in Berlin, working there as a doctor of economics. He is presumed to have passed away there in the 1960s.

Max Schick

He was released from French captivity in February 1920 and returned to his hometown Munich. During the 2nd World War, he was promoted Hauptmann (Reserve) of the anti-aircraft artillery. Further information concerning his life has not been found by the authors.

Ludwig Schmid

After being wounded severely, he underwent treatment in various hospitals, and saw the end of the war in the homeland. Over the turn of the year 1918/19, he was sent on leave to Ansbach and Nuremberg, and was transferred to his parent unit, the 8th Bavarian Field Artillery Regiment. On March 31, 1920 his military service ended. Ludwig Schmid died in Munich on December 26, 1926, due to long-term complications of his wounding.

Maximilian Veith

After the end of the war and demobilization of Jagdstaffel 76 for the Bavarian 3rd Infantry regiment back and there as an adjutant at the 3rd Company. From April 1, 1919 he served as an adjutant in the Detachement of Lütz and the Iron Division and was involved in various battles in the Baltic States.

On March 16, 1920, he departed the Verband der Eisernen Division (Association of the Iron Division) and joined the disbanding office of the Bavarian 3rd Infantry Regiment. He was discharged from military service on the last day of April 1920. Subsequently, he found employed at the Army Processing Office Bavaria. On June 7, 1921 he was appointed char. Lieutenant (retired). His further active service with the new Wehrmacht can only be proven in 1934, by then he was serving with Infantry Regiment 41. He was appointed Oberst (Colonel) on New Year´s Day 1943. On January 10, 1943, while serving as commander of Grenadier Regiment 670, he fell during the battle for Stalingrad.

Clemens Vollert

After his release from military service, Clemens Vollert turned his hobby, art painting, into a profession. He married his fiancée Anna and lived and worked as an artist in Munich. He died there on July 16, 1956 and found his last rest in the East Cemetery in Munich.

Siegfried Walther

On June 21, 1918, he was transferred to Palestine to Armee-Flugpark F and assigned to Jagdstaffel F. As already mentioned, he was killed in combat there. Reports on the exact date and circumstances of his death differ in detail.

His wartime service records indicate the August 24, 1918 as the date of his death. Norbert Schwake's book, „Deutsche Soldatengräber in Israel" (German Soldier Tombs in Israel), also refers to this date. In addition, the book notes that his aircraft was set on fire over Tul Kam (today Tulkarm). Reportedly, he took to his parachute, but it apparently failed to deploy, and he fatally fell at Kalkilya (today's Qalqilya).

In slight contrast to this, the „Verlustliste der deutschen Luftstreitkräfte im Weltkrieg" (Loss List of the German Aerial Forces in the World War), published in 1930, indicates the date of his death as August 25, 1918, as well as his tomb being located at the military cemetery of Nazareth.

The crew of Lt. P. J. McGinnes and Lt. H. B. Fletcher, flying with 1. Sqn. Australian Flying Corps claimed shooting down two German fighter planes in the aforementioned area on August 24, 1918. Should this be the true date of his death this crew could have been Siegfried Walther's opponents over Tul Karam.

Moritz Weig

After his release from English captivity, Moritz Weig returned back home on February 14, 1920 via the Dulag (transit camp) Emden. Later, he worked as a merchant in Stuttgart. He was reactivated to the Luftwaffe, probably at the beginning of the war in 1939. On November 17, 1942 he was promoted to Captain (z.V.), a command to the Battalion Leader School in Antwerp later followed. On March 1, 1943 he was appointed Major of the Reserve. Further information is not known by the authors.

Erwin Wenig

After his transfer from Jagdstaffel 76 to Flieger-Ersatz-Abteilung 1 in Schleißheim, Erwin Wenig was appointed squadron leader of the newly formed Jagdstaffel 80 on January 25, 1918. He now had to face the task of preparing the squadron for its imminent frontline service. At the end of February 1918, the Staffel was moved to its first airfield located near the small village of Morsberg, in the area of the 19th Army. On June 18, 1918, he was slightly wounded in aerial combat in the right foot by a bullet, but he remained with the Staffel.

After the end of the war, he was transferred to FEA 1b Schleißheim on January 14, 1919 appointed leader of the Alert Staffeln. During the riots in and around Munich, he registered for the Voluntary Fliegerabteilung at the end of April 1919, which was renamed Truppenfliegerstaffel (Troop Pilot Squadron) 21b after the riots. On May 25, 1919 he was transferred back to Würzburg to join the parent unit of the 2nd News battalion. On February 15, 1920 Erwin Wenig joined the disbanding office of the 8th Bavarian Field Artillery Regiment. Afterwards, he returned into civilian life, and turned his hobby, photography, into a profession by founding a photo shop in his hometown.

In the 1930s, Erwin Wenig enlisted for the Luftwaffe again. Unfortunately, only fragmentary information concerning his service has been found. It is known that on 1 July 1935 was promoted to Major and he served in 1939 in Lechfeld at the Great Fighter Pilot School there. On April 1, 1940, the fighter pilot school was moved to Warsaw-Okecie and renamed Große Kampffliegerschule 3. Major Wenig was given the position of commander the 1st Group. At the beginning of 1943, he took over the leadership of the Air Service Command 1/6 in Münster-Loddernheide.

After the end of the 2nd World War, he returned to his birthplace Weiler and ran a photo shop until his retirement. He passed away there on March 5, 1960.

15 One Century after the War

by Philippe Charroin
Aéroplane Restauration Management, Le Bourget, France

(All photos of this chapter, unless otherwise indicated, are from the Aéroplane Restauration Management.)

In 2005, Memorial Flight was visited by two elderly gentlemen claiming to have dug up the substantial remains of a World War I aircraft from a swamp on their property near Montdidier, in the Somme region.

Claims like the ones made by the two visitors are pretty common, but upon closer examination of such claims the usual outcome ranges from tractor parts to boat anchors! However, what these gentlemen had was far more interesting, with not only a lot of splintered wood but an almost complete Mercedes engine and mechanical appliances typically found on a German WW I aircraft, all in remarkably good condition. The engine sustained a lot of damage in the crash, but a few of its components are in good enough condition to be used again. The fact that the aeroplane fell into a patch of boggy ground meant that many the wooden and metal parts survived the passage of almost a century surprisingly well. Of course, a full restoration of the surviving components into an airworthy aircraft will not be possible, but components that are found to be in good enough condition will be re-used.

Present amongst the selection of retrieved parts was the battered remain of the data plate identifying the aircraft as an Albatros D Va carrying the factory serial number 4360. This serial number places the airplane in a batch of 250 aircraft, ordered from the Albatros company in Berlin-Johannisthal, which were assigned the military order numbers 5600/17 to 5849/17. The serial number of the Mercedes D IIIa engine is 35317, and it came off the production line on November 15, 1917.

This recovery is as close to a miracle as it can be. The two gentlemen knew the airplane was there because a relative saw it coming down when he was a kid in the summer of 1918, and repeatedly told this story to them. When land works were carried out on their property, they seized the opportunity and borrowed a huge excavator to dig exactly at the location pointed out by their relative, and there it was! Without the eyewitness account, and the clear memory of the older related, there would have been absolutely no chance the aircraft remains could have been locacted otherwise.To our knowledge, this is the only documented case of the recovery of an untouched, original WW I fighter aircraft in modern time.

The examination of all the remains revealed that at least 70% of the airplane was there, representing a tremendous amount of data, especially the front section up to the pilot seat. One should keep in mind that 15 years ago, the thought of being able to rebuild an accurate Albatros D Va was pure fantasy. No original drawings were known to exist and only two original aircraft survived, one in Washington, USA, and the other one in Canberra, Australia. The only data then available was presented in a book published by the Smithsonian Institute following the restoration of their Albatros D Va, which also included drawings of the type made by Bob Waugh, measured on the second surviving original Albatros in Canberra. However, these were known to feature some mistakes. Both aircraft could, of course, still be used for external reference, but there was no way anyone could cut open the structure or the tanks to document components hidden inside the airframe. Even disassembling mechanical appliances like the fuel distribution system was not an option.

This re-discovered aircraft, fallen from the sky, or coming back up from the ground, presented a golden opportunity to access new data on the basis of an in-depth research, with the aim of producing a new set of construction drawings. This huge task has been an international team effort involving both a partner in California as well as another one in New Zealand. It involved visits to Washington, USA, Canberra in Australia, and to a few private collectors around the world who own original D Va components.

Measurements were taken from the Canberra aeroplane which was luckily undergoing restoration at that time. In exchange, a small group of memorial flight members flew down under to assist in the re-application of fabric to the wings by teaching the museum´s restoration staff the art of WWI German fabric covering practices. The New Zealand team was able to digitize the entire fuselage of this same aircraft by using the latest laser scanner.

Also, via various sources, we learned about the existence of someone who had devoted a considerable amount of time to taking pictures and measurements of the NASM's Albatros in Washington during her restoration in the late 70´s. Two or three of his sketches could be found on a few websites, and it was said he had a huge amount of data at this disposal. It was worth the long quest to locate and finally meet Frank Garove in Baltimore, and the sketches and pictures he shared with us have been a major aid in the successful completion of the project. This work gave birth to a 3D model of the Albatros and a new set of 2D fabrication drawings, which led to the production of a series of Albatros DVa by "The Vintage Aviator" in New Zealand.

Gathering the required information and actually making these drawings took a very long time, during which we continued our research in order to discover the identity and history of our airplane. From the outset, we knew the airplane crashed during the summer 1918. Another interesting point worth noting was the fact that were there were no human remains to be found, and the seat belt was unfastened. The engine controls, all of which were recovered, were set in the emergency positions. The seat also had the remains of the old seat cushion attachment cut off under the seat where the cushion had been replaced by the parachute, which had apparently been used *on the last flight of this aircraft*.

In August 2011, two unfinished airframes arrived at the Memorial Workshop near Paris, coming from New Zealand, in order to be completed to flying condition, with one intended to be completed for the partner in California. As of today, the two projects are almost completed. The Memorial Flight's aircraft is already painted in Bavarian blue and white stripes and will represent, but not claim to be, 5765/17, the aircraft abandoned by Lt. Schick, and which supplied so much information. Quite a few restored original parts have already found their way into the reproduction, such as the fuel gauge or the air pump. Some of the instrumentation, for example the altimeter, has only been restored cosmetically, since the aeroplane will only fly locally around the airfield, so the altimeter is not required to actually function as intended.

Once finished, the aeroplane will join our WWI fleet in La Ferté Alais and, flying alongside our original SPAD XIII, will pay tribute to all the WWI aviators, and keep the memory of the Bavarian Jagdstaffel alive!

Above: The swamp mear Bousicourt where the aircraft was found.

Above: Members of the Aeroplane Restoration Management in front of various components of Albatros D Va. D.5767/17 spread out after their recovery.

Left: The main fuel tank suffered badly during the crash, but much of it could be recovered.

Above: The placard square placard mounted to the tank gave the capacity of 78 litres, type of aircraft and serial number of the tank. Below it, the round "Z.A.K." placard indicated the item was checked and accepted by the central acceptance commission.

Above: View of one of the machine guns with the cooling jacket being badly bent by the impact.

Left: The multi-purpose control valve, with fuel filter barely visibly at the bottom of the photo, survived in remarkably good condition.

Right: A thick layer of rust covered the cylinders of the Mercedes D.IIIa engine.

Below: The recovered Albatros manufacturer´s plate indicate that the aircraft was manufactured in Berlin-Johannisthal, was a D Va and carried the factory serial number 4360, which was different from the military number.

Höhe in km
D.R.P.a.
N.S.
Nr. 13218
0
1
2
3
4
5

Above: Fuselage of the faithfully reconstructed aircraft.

Facing Page, Above: The magnetic compass complete with its mounting.

Facing Page, Below: The scale of the altimeter fitted to the aircraft went up to 6.000 meters.

Right: View in the Cockpit of the Albatros D Va.

Above: View into the cockpit of the reconstructed D.5767/17. Switches and levers have already been fitted, including the fuel clock.

Above: Components of the crashed Albatros D Va D.5765/17 retrieved from the swamp near Montdidier could be identified and assigned to the aircraft of Jagdstaffel 76.

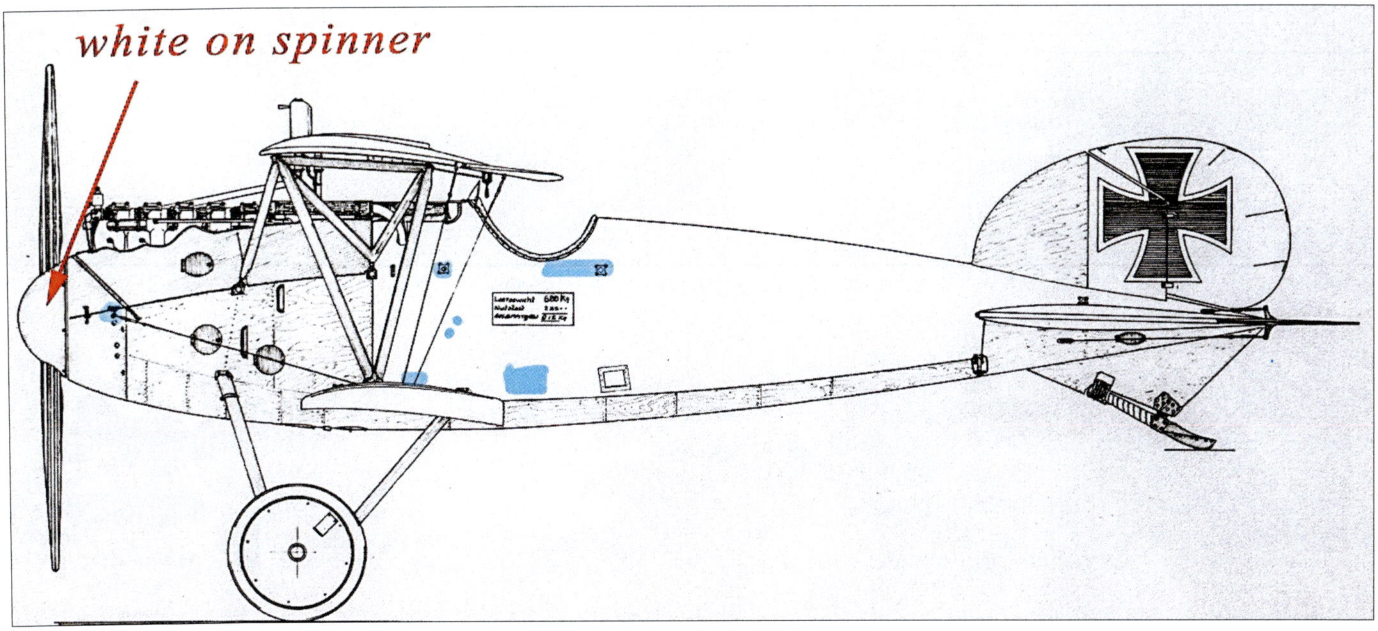

Above: The drawing shows the blue pieces of wood found on the Albatros, which identifies the aircraft as the white and blue striped aircraft, finally identified as one of the aircraft of Lt Bärenfänger.

Below: Eventually, two flare cartridge holders have been installed, and the fuselage paint job has left out the Albatros manufacturer´s plate on the nose of the aircraft.

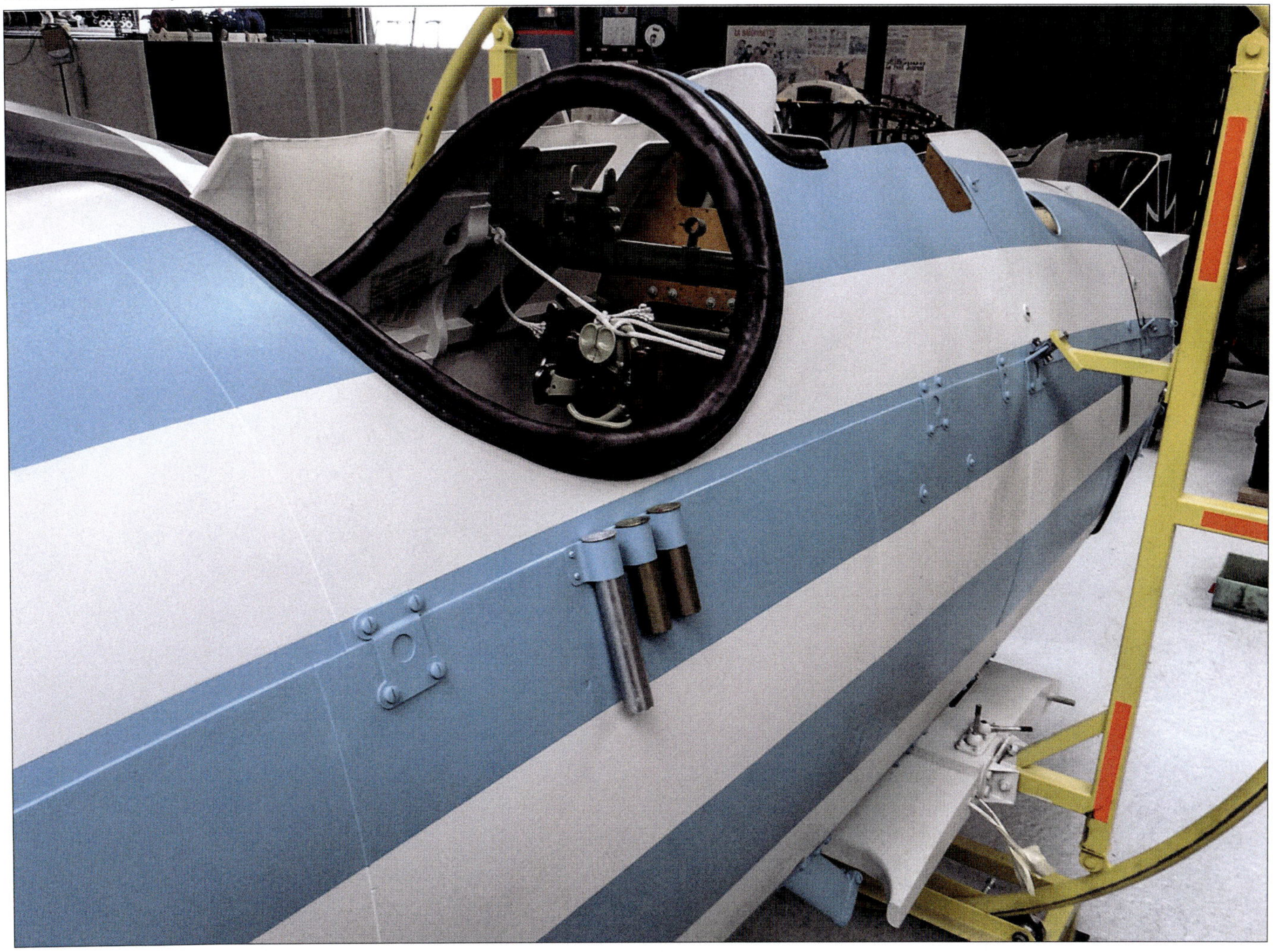

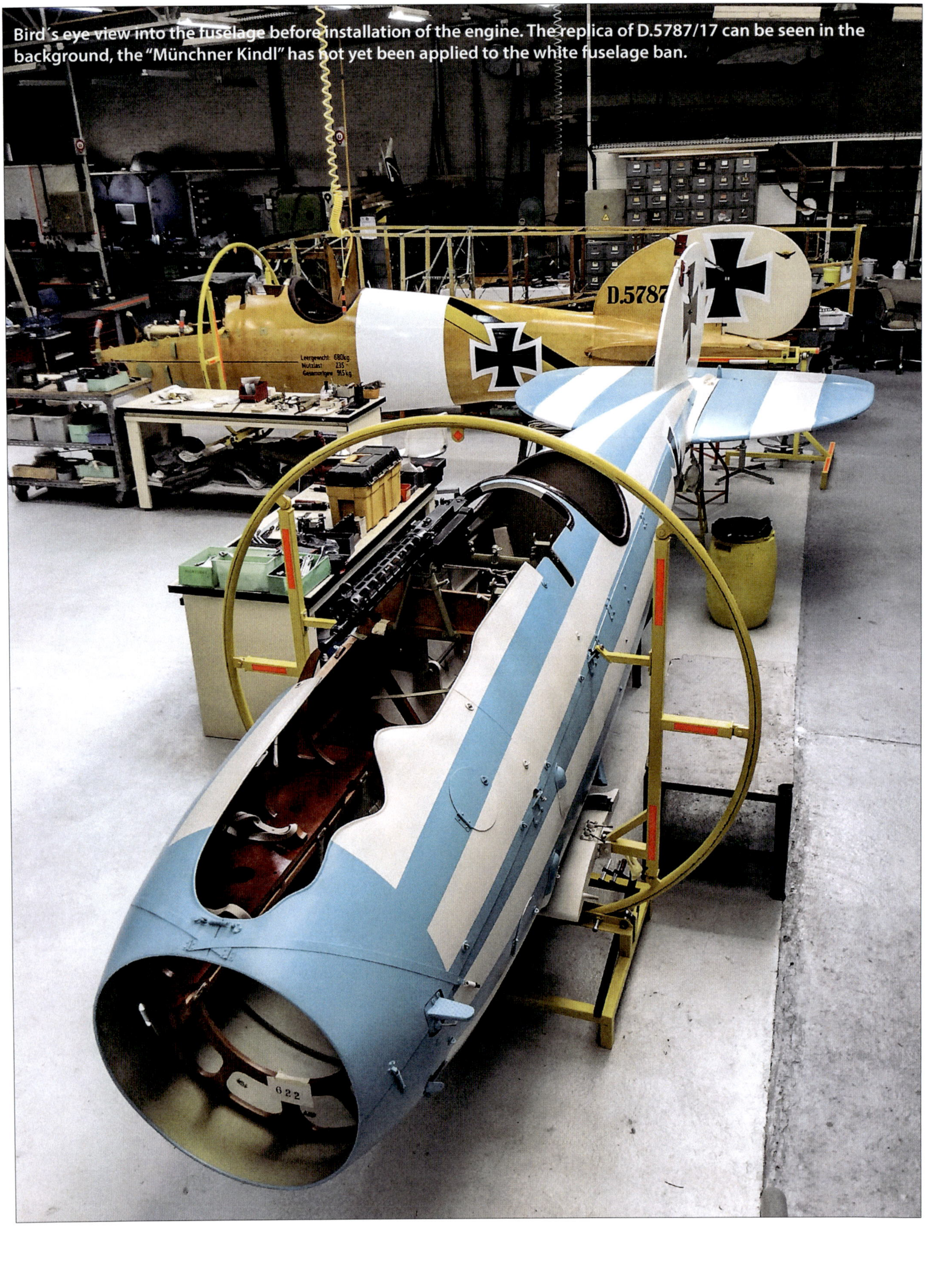

Bird´s eye view into the fuselage before installation of the engine. The replica of D.5787/17 can be seen in the background, the “Münchner Kindl” has not yet been applied to the white fuselage ban.

The same shade of blue used for the fuselage stripes has been applied to the horizontal tailplane.

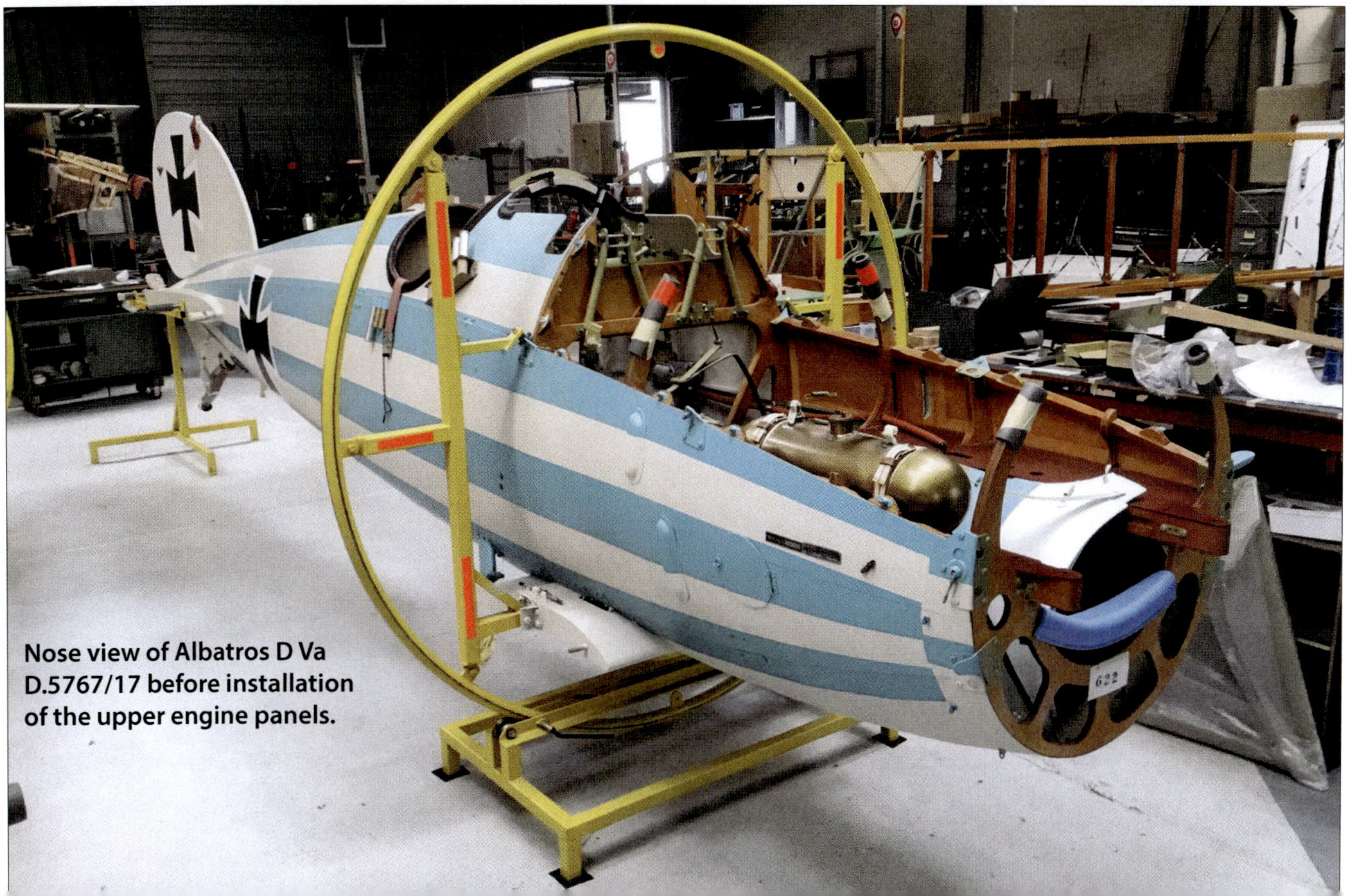

Nose view of Albatros D Va D.5767/17 before installation of the upper engine panels.

Left: Albatros D Va D.5787/17 was constructed alongside its sister aircraft.

Facing Page, Above: Albatros D Va D.5787/17 "Münchner Kindl" under construction.

Facing Page, Below: Pattern drawing of the "Münchner Kindl" marking that was eventually applied to the white fuselage band of D.5787/17.

Below: A view of both fuselages alongside allows comparison of the different flare cartridge holders. Note the protruding tubes onto which the horizontal tailplane was mounted on the nearest aircraft.

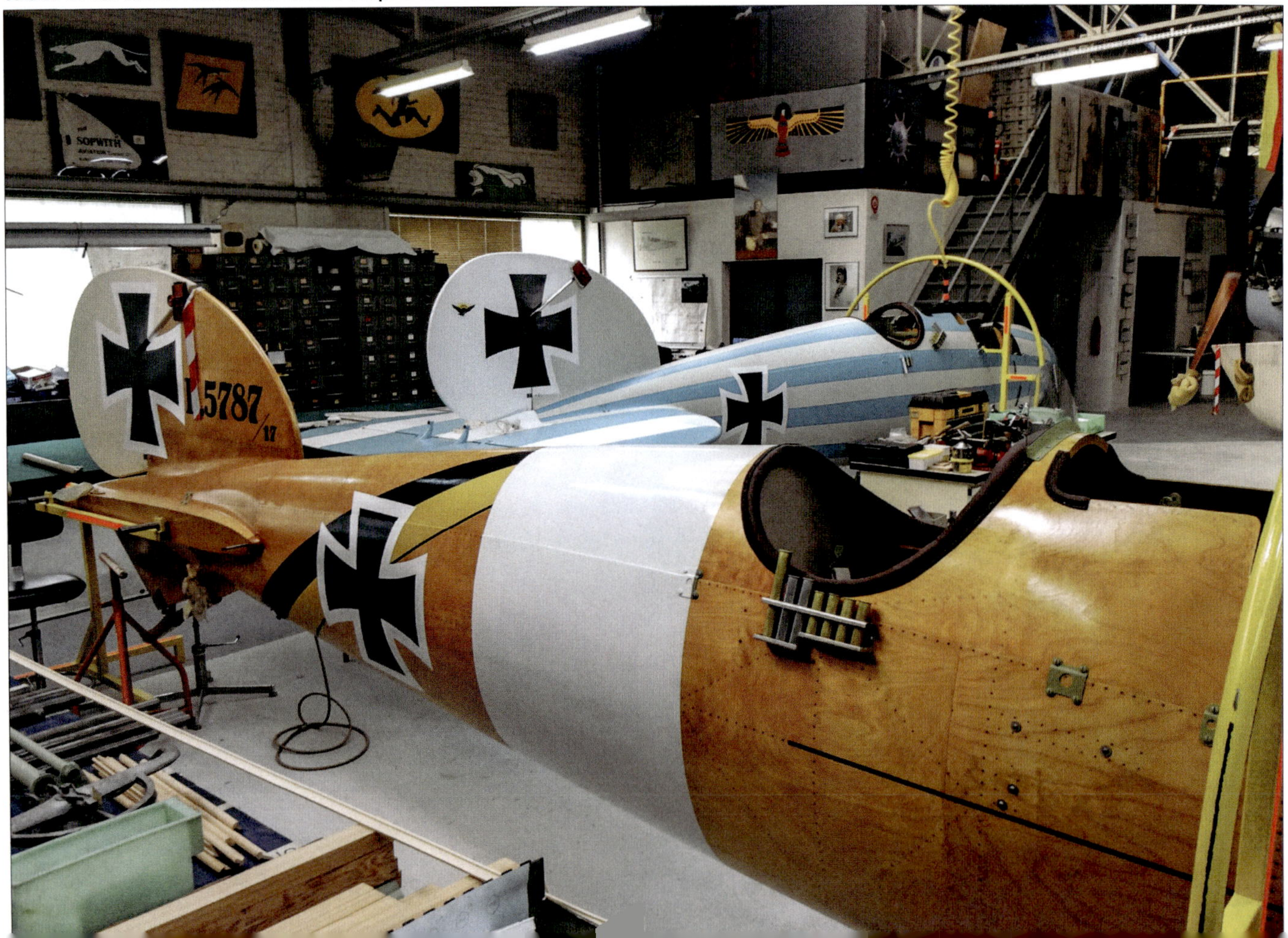

D.5787
Leergewicht: 680kg.
Nutzlast: 235
Gesammtgew: 915kg

wicht: 680kg.
st: 235
mtgew: 915kg.
HB

Above: This Albatros D Va replica was built by Cole Palen, Old Rhinebeck Aerodrome and his team in the 1970´s. (A. King)

Below: In 2014, the aircraft was fully restored to flying condition and given a new paintjob, replicating the Jasta 76 aircraft flown by Lt. Hans Böhning in early 1918. (A. King)

Above: A very good impression of what the original Albatros appeared like to its opponents is offered in this fine in-flight view. (T. Polapink/M. Beaumont)

Below: Another realistic in-flight view over a landscape remarkably similar to the Alsace region where Jagdstaffel 76 operated in early 1918. (T. Polapink/M. Beaumont)

16. Endnotes

Prologue: Jagdstaffel 19, St. Loupe en Champagne, 7th German Army

1. Otto Fuchs' personal remarks to Bruno Schmäling in 1978.
2. Letter from Walter Böning, October 6, 1917. Private ownership, Archive copy, Bruno Schmäling.
3. Personal information shared with Bruno Schmäling by Dr. Walter Böning in 1977

1. Formation of Royal Bavarian Jagdstaffel 76 Schleißheim, September 1917 – November 1917

1. Personnel File OP 38021 Walter Böning, Bavarian Central State Archive, Department 4, Munich. Aktiver Offizier and Reserveoffizier designations are elaborated on in the brief biographies of each respective Staffel member. They are omitted in the remainder of the text as these designations were frequently not listed, even in official documents. Furthermore, frequent repetition of the designations would be tiring for the reader.
2. Personal information provided by Dr. Walter Böning in conversation with Bruno Schmäling, 1977
3. OP 336 August Ertl, ILuft 204, Bavarian Central State Archive, Department 4, Munich
4. Letter from Walter Böning, July 27, 1916
5. Letter from Walter Böning, July 30, 1916
6. Letter from Walter Böning, September 21, 1916
7. Letter from Walter Böning, September 25, 1916. "Bloh" was a peasantry located to the north-west of Oldenburg, it is now part of the city of Bad Zwischenahn. The creek "Ofener Bäke" runs nearby, and consequently there were lots of mosquitos buzzing through the air in the summer.
8. Letter by Walter Böning, September 25, 1916
9. Letter by Walter Böning, October 2, 1916
10. Letter by Walter Böning, October 30, 1916
11. Letter by Walter Böning, November 11, 1916
12. Confirmation by Kogenluft 103 666, April 24, 1917, archive Bruno Schmäling. OP 38021 Walter Böning, Bavarian Central State Archive, Department 4, Munich.
    According to the war diary excerpt by Erich Tornuß the French aircraft was a Caudron (presumably an older variant, possibly a G.III, which would fit the description).
13. Personal information by Dr. Walter Böning, ibid.
14. Personnel file OP 38021 Walter Böning, Assessment by Lt. Ernst Hess, Bavarian Central State Archive, Department 4, Munich
15. This is a reference to the following aerial victories:
    April 4, 1917 Gitterschwanz (Lattice tail aircraft)
    April 30, 1917 Captive Balloon
    April 29, 1917 Caudron
    August 17, 1917 Caudron
    September 23, 1917 two Nieuport single-seater aircraft
16. Letter from Walter Böning, October 11, 1917
    I-Luft: Bayerische Inspektion des Militärluftfahrwesens (Bavarian inspectorate of military aviation)
17. Letter from Walter Böning, October 20, 1917.
18. Personal information by Frau Vollert, widow of Clemens Vollert, shared with Bruno Schmäling, 1980.
19. Personnel file OP 17694 Clemens Vollert, Bavarian Central State Archive, Department 4, Munich
20. Letter from Walter Böning, November 4, 1917, ibid.
21. Personal information by Dr. Walter Böning ibid.
22. Personal information by Dr. Walter Böning ibid.
23. Letter from Walter Böning, October 27, 1917, ibid. In Flanders, German pilots mostly faced British units which had been ordered to use "offensive strategies" to cross enemy lines at every opportunity in order to face German aircraft overhead their own territory. This aggressive way of fighting significantly increased the pressure on German Fliegertruppen; on the British side, it led to substantial losses.
24. Letter from Walter Böning, November 4, 1917, ibid.

2. Baptism of Fire in Alsace Armee-Abteilung B, November 1917 – January 1918

1. Kriegstagebuch (KTB, war diary) transcript Jagdstaffel 41, Erich Tornuß, Winfried Bock collection.
2. Letter from Walter Böning, November 4, 1917, ibid.
3. Letter from Walter Böning, November 9, 1917, ibid.
4. Handwritten note about personal information given to Alex Imrie by Joachim von Hippel, Imrie Collection.

5. Personnel file OP 42943 Wilhelm Kolb, Bavarian Main State Archive Munich, Department 4
6. Personal information Dr. Walter Böning ibid.
7. Kriegsstammrolle (KrStR, Personnel Roster) 18157/2554 Siegfried Walther, Bavarian Main State Archive Munich, Department 4
8. Personal information Dr. Walter Böning ibid.
9. Unpublished text about the story of the live of Hans Böhning as well as personal information by Frau Agnes Schwörer, widow of Hans Böhning, to Bruno Schmäling. The statement made in the "Jasta Pilots" book by Norman Franks and Frank Bailey, indicating that Grottau/Reichenberg was located in the Kingdom of Bavaria, is incorrect.
10. Unpublished text about the story of the live of Hans Böhning, archive Bruno Schmäling.
11. Personnel file OP 5021 Hans Böhning, Bavarian Main State Archive Munich, Department 4. Technischer Offizier (technical officer) was not an official rank designation used by the Fliegertruppe. Rather, it described a role taken on by a pilot who had some technical training or specific interests in technical aspects. The duties of a "Technischer Offizier" were unofficially added to the respective pilots' flying duties
12. Letter form Walter Böning, November 27, 1917, ibid.
13. Personal information by Dr. Walter Böning ibid.
14. Otto Kissenberth later became a successful fighter pilot with Jagdstaffel 16 and Staffelführer of Jagdstaffel 23. He was also decorated with the "Pour le Mérite" on July 24, 1918 while he was hospitalized, recovering from injuries suffered during a crash after a test flight in a captured Sopwith Camel about eight weeks earlier.
15. Personnel file OP 29523 Erwin Wenig, Bavarian Main State Archive Munich, Department 4
16. Personal information and correspondence Oberst a.D. Max Speidel to Bruno Schmäling, personal information by Frau Wenig to Bruno Schmäling.
17. Letter from Walter Böning, November 9, 1917, ibid.
18. Personnel file OP 6006 Franz-Xaver Danhuber, Bavarian Main State Archive Munich, Department 4.
19. Personnel file OP 6166 Karolus Bärenfänger, Bavarian Main State Archive Munich, Department 4, assessment of Bärenfänger by Lt. Schäfer, Jasta 28, Kastner archives. Kastner, Reinhard, Compilation of personnel files, ibid.
20. KrStR 17987/16; 18018/147, Paul Wagner, Bavarian Main State Archive Munich, Dept. 4
21. Letter from Walter Böning, November 27, 1917, ibid.
22. Letter from Walter Böning, November 27, 1917, ibid.
23. Personal information Dr. Walter Böning ibid.
24. Personal information Dr. Walter Böning ibid.
25. According to coinciding statements by former fighter pilots such as Theodor Rumpel, Walter Böning, and Otto Fuchs, French anti-aircraft artillery units were well-trained and considered to fire fairly accurately.
26. Résumé des opérations aériennes, collection Winfried Bock
27. Report of the Kommandeur der Flieger der Armee-Abteilung B, Copy collection Winfried Bock.
28. Letter from Walter Böning, December 15, 1917, ibid.
29. Personnel file OP 42943 Kolb, Bavarian Main State Archive Munich, Dept. 4
30. Letter from Walter Böning, December 15, 1917, ibid.
31. Personal information Dr. Walter Böning ibid.
32. KrStR 17994/190; 18026; 18018, Karl Hopf, Bavarian Main State Archive Munich, Dept. 4
33. Personal information Oberst. a. D. Max Speidel, ibid.
34. Letter from Walter Böning, January 2, 1918, ibid.
35. Letter from Walter Böning, January 11, 1918, ibid.
36. Letter from Walter Böning, January 16, 1918, ibid.
37. Personal information Dr. Walter Böning ibid.

3. Aerial Combat over the Vosges Armee-Abteilung B, February 1918 – March 1918

1. Letter from Walter Böning, February 2, 1918, ibid.
2. Résumé des opérations aériennes, ibid.
3. Letter from Walter Böning, February 7, 1918, ibid.
4. According to "Nachrichtenblatt", the victory claimed on February 4, 1918 was confirmed as the 8th aerial victory, and the February 11 victory as the 9th. This book lists the aerial victories in accordance with the recognition list.
5. Résumé des opérations aériennes, ibid.
6. Letter from Walter Böning, February 14, 1918, ibid.
7. Résumé des opérations aériennes, ibid.
8. KrStR 18018/143, Josef Diem, Bavarian Main State Archive Munich, Department 4
9. Personal information, Dr. Walter Böning ibid.
10. KrStL 18016/7, Hans Hertrich, Bavarian Main State Archive Munich, Department 4

11. KrStR 18022/8; 18098/994, Xaver Hüttner, Bavarian Main State Archive Munich, Department 4
12. Letter from Walter Böning, March 7, 1918, ibid. Winfried Bock collection
13. Letter from Walter Böning, March 7, 1918, ibid.
14. Résumé des opérations aériennes, ibid.
15. According to Erich Tornuß the war diary was not held in the archive in Potsdam. Therefore, only fragmented information is available. What we do know is that it was in Habsheim on March 10.
16. Letter from Walter Böning, March 13, 1918, ibid.
17. In order for an aerial victory to be officially recognized by the German authorities, confirmation provided by independent ground witnesses was mandatory. Statements provided by fellow pilots, as was the case in the British RFC/RAF, were not regarded as sufficient to obtain confirmation for a victory claim. Several aerial victories that occurred far across enemy lines were therefore not given official confirmation.
18. Letter from Walter Böning, March 16, 1918, ibid.
19. Résumé des opérations aériennes, ibid.

4. The British- a new Opponent 2nd German Army, March 19, 1918- July 9, 1918

4.1 Liéramont Airfield March 20, 1918 – April 15, 1918

1. Personnel file OP 33450, Richard Emmerich, Bavarian Main State Archive, Department 4, Munich
2. KrStR 18018/153; 18209/279; 18976/49, Gottfried Stemmler, Bavarian Main State Archive, Department 4, Munich
3. Letter from Walter Böning, April 3, 1918, ibid.
4. Personnel file OP 41881 Hans Hertrich, Bavarian Main State Archive, Department 4, Munich
5. KrStR 18144/29; 18016/9, Walter Angermund, Bavarian Main State Archive, Department 4, Munich
6. KrStR 18003/422; 18098/706; 18018/152, Karl Koller, Bavarian Main State Archive, Department 4, Munich
7. Letter from Walter Böning, April 12, 1918, ibid.
8. Personnel file OP 33450 Richard Emmerich, Bavarian Main State Archive, Department 4, Munich
9. Combat reports 15. Sqn. RAF, summarized in the daily intelligence summary of the RAF („War Diary"), Winfried Bock collection
10. War diary transcript, Erich Tornuß, ibid. Winfried Bock collection.
11. Combat report 56. Sqn. RAF, ibid.
12. Letter from Walter Böning, April 12, 1918, ibid.
13. Personal information by Dr. Walter Böning ibid.
14. KrStR 18018/155; 17943/424, Ludwig Artmann, Bavarian Main State Archive, Department 4, Munich
15. Letter from Walter Böning, April 14, 1918, ibid.
16. Personal information by Dr. Walter Böning ibid.

4.2 Airfield Suzanne April 15, 1918 – July 9, 1918

1. Personal File, Siegfried Walther, KrStR 18018/106, Bavarian Main State Archive, Department 4, Munich
2. Personal information by Dr. Walter Böning ibid.
3. Personal information by Dr. Walter Böning ibid.
4. KrStL Moritz Weig, 18016/10, Bavarian Main State Archive, Department 4, Munich
5. Personnel file Karl Edmund Neu, OP 9084, war personnel roster 18079/13; 18016, Bavarian Main State Archive, Department 4, Munich
6. Personnel file, Hermann Obermeier, OP 27232, KrStL 18079/40, 18016/12, Bavarian Main State Archive, Department 4, Munich
7. Personal information by Dr. Walter Böning ibid.
8. Capt. Junor was not shot down until the late afternoon; according to Jagdgeschwader 1 war diary entries, Lt. Koepsch shot down an S.E. 5 at "6.55 in the morning". All other war diary entries on that day relating victories and losses (emergency landings) that day also occurred in the morning hours only. Jasta 4 flew at this time Fokker Dr.I
9. Letter from Walter Böning, April 20, 1918, ibid.
10. Alex Revell, High in the Empty Blue. The History of 56. Squadron, RFC/RAF 1916 to 1920. Page 289 and 290, Flying Machine Press 1995
11. http://pilotesdechasse.over-blog.com/article-coadou-marcel-100684007.html
12. KrStR Josef Diem, 18018/143, Bavarian Main State Archive, Department 4, Munich
13. KrStR Franz Martin, KrStL 18016/14, Bavarian Main State Archive, Department 4, Munich
14. Personnel file, Georg Markert, 18018/160; KrStR 17963/3426. Bavarian Main State Archive, Department 4, Munich
15. KrStL Max Schick, 18016/13, Bavarian Main State Archive, Department 4, Munich
16. Letter from Walter Böning, May 21, 1918, ibid.
17. Intelligence Summary RAF, Report Capt. J.D. Belgrave 60. Sqn. R.A.F., ibid.
18. Letter from Walter Böning, May 24, 1918, ibid.
19. Intelligence Summary RAF, Report 65. Sqn.

R.A.F, ibid.
20. Letter from Karl Koller written from British captivity on July 7, 1918. B. Schmäling Archives
21. Personal information by Dr. Walter Böning ibid.
22. Letter from Georg Markert to his wife, May 30, 1918; R. Kastner Archive copy
23. Letter from Ludwig Altmann, June 2, 1918 (R. Kastner Archive copy)
24. Norman Franks, Frank Bailey & Rick Duiven, The Jasta Pilots, London 1996
25. Personal information by Dr. Walter Böning ibid.
26. Letter from Walter Böning, June 1, 1918, ibid.
27. Letter from Walter Böning, June 5, 1918, ibid.
28. Personal information by Dr. Walter Böning ibid.
29. Letter from Walter Böning, June 23, 1918, ibid.
30. Personal information by Dr. Walter Böning ibid.
31. KrStL Maximilian Veith, 18016/15, Bavarian Main State Archive, Munich
32. Personnel file, OP Karl Dumler 3319, Bavarian Main State Archive, Munich
33. Personnel file, OP 3319 Karl Dumler, ibid.
34. KrStL 18016/6 Karolus Bärenfänger, Bavarian Main State Archive, Munich
35. Personal information by Dr. Walter Böning ibid.
36. Personnel file, OP Ludwig Schmid, 48908, KrstR. 18031/108, Bavarian Main State Archive, Munich
37. War diary excerpt Jagdstaffel 34, transcription by Erich Tornuß with correction and supplementation by Dr. Gustav Bock, Winfried Bock.
38. Personal information by Heinz Arntzen, Jasta 15 and Staffelführer Jasta 50 in conversation with Bruno Schmäling
39. Personal information by Heinz Arntzen and Theodor Rumpel, Jasta 23, ibid.
40. KrstR. Josef Reinhard, 18018/170, , Bavarian Main State Archive, Department 4, Munich
41. Krst.R. Johann Brunner 18081/703; 18125/379, Bavarian Main State Archive, Department 4, Munich
42. Personnel file, Walter Curtius, OP 5914 KrStL 18144/210; 18016/16, Bavarian Main State Archive, Munich
43. Personnel file, Walter Curtius, OP 5914 KrStL 18144/210; 18016/16, Bavarian Main State Archive, Munich

5. Offensive Battle at the Marne and Champagne, 1st German Army, July 9, 1918 – July 17, 1918

1. Personal information by Mrs. Luise Wildmann, daughter of Josef Bauer to Bruno Schmäling.
2. KrStR 18018/, Josef Bauer, Bavarian Main State Archive, Munich.
3. KrStR 18018/175, Julius Geyer, Bavarian Main State Archive, Munich
4. KrStR 18018/176; 17943/425, Rupert Hainzl, Bavarian Main State Archive, Munich
5. KrStR 18018/177; 18210/19, Erwin Hepperlin, Bavarian Main State Archive, Munich
6. Personnel file, OP 47359 Amandus Rostock, Bavarian Main State Archive, Munich
7. Kriegsstammrolle der Jagdstaffel 76, Bavarian Main State Archive, Munich
8. Events recalled by Josef Bauer, Private property, Transcript B. Schmäling Archive
9. Lt. Fritz von Borries' logbook, Jagdstaffel 22. According to his logbook entries, Jasta 22 had been equipped with Fokker D VII aircraft in July 1918.
   According to the photos and documents in the collections of Hermann Leptien and Hermann Böhler, Jagdstaffel 63 had begun taking delivery of the Fokker D VII in early August 1918, all photos and documents archive Bruno Schmäling
10. War diary excerpt Jagdstaffel 63 transcription by Erich Tornuß, copy archive Bruno Schmäling
11. Minutes recorded by Josef Bauer, personell Information by Frau Wuldmann, daughter of Josef Bauer to Bruno Schmäling.

6. Back at the old Base in Upper Alsace German Armee-Abteilung B, August 20, 1918 until the End of War

1. Photographs and documents of Julius Geyer, photographs from the collections of Greg VanWyngarden, Rainer Absmeier and Jörn Leckscheid
2. OP 41881 Hans Hertrich, Bavarian Main State Archive, Department 4, Munich
3. Photographs and documents of Julius Geyer, photographs from the collections of Greg VanWyngarden, Rainer Absmeier, and Jörn Leckscheid
4. Letter from Julius Geyer, September 20, 1918, transcripts archive Bruno Schmäling
5. Jagdstaffel 76 ranking list, Bavarian Main State Archive, Munich
6. Personnel file Amandus Rostock OP 47359, Bavarian Main State Archive, Munich. According to the Jagdstaffcl 76 ranking list, the Staffel relocated on May 3, 1918.
7. Personnel file Amandus Rostock, ibid.
8. Personal information by Otto Fuchs, Jagdstaffel 77 in conversation with Bruno Schmäling
9. Personal information by Joachim von Hippel, Jagdstaffel 71 in conversation with Alex

Imrie
10. Résumé des opérations aériennes, compilation of French reports, Winfried Bock collection.
11. Personal information by Walter Böning, ibid.
12. Letter by Walter Böning, November 11, 1918. If Walter Böning´s expression would be taken literally, Oblt. Rostock would have arrived at Jagdstaffel 76 on October 23, 1918. However, the Jagdstaffel 77 war diary transcript by Erich Tornuß proves that he was transferred as the new commander of Jasta 76 on October 7, 1918, and actually arrived at his new unit on that day. For this reason, Bönings claim cannot be true. Supposedly, he was trying to express that Rostock had only recently taken command of the unit, hence the choice of wording "8 days ago".
13. Bericht des Kommandeurs der Flieger Armee-Abteilung B, November 1918, 14.Letter from Walter Böning, November 11, 1918, ibid.
14. Letter from Walter Böning, November 11, 1918, ibid.
15. Personal information by Walter Böning, ibid. A number of Bavarian Jagdstaffeln, including Jagdgeschwader IV and Jagdstaffel 78, flew their aircraft, or some of their aircraft, back to Germany.

7. Aircraft Markings and Paint Schemes

1. The color "red" was not used exclusively as a marking on aircraft by Jasta 11 or Manfred von Richthofen. A number of pilots flying with other units also used red as their personal marking, often as a reference to their parent unit, such as the infantry or a certain cavalry regiment.
2. Personal information provided by Walter Böining and the widow of Hans Böhning, Mrs. Agnes Schwörer, to Bruno Schmäling.
3. Information of the Hopf family to Bruno Schmäling.
4. G-Report File No. G/3Brigade7, Archive G. VanWyngarden.
5. Research from the Aeroplane Restauration Management, Le Bourget, France.
6. Documents of Julius Geyer provided by the family to Bruno Schmäling.
7. According to the note written on the back of the photo of his Fokker D VII, he used a white ribbon as personal marking. See footnote to Profile 32.
8. Information of Otto Fuchs to Bruno Schmäling and photos in his photo album.

Bibliography

Publications

Franks, Norman, Bailey, Frank & Duiven, Rick, The Jasta Pilots, London 1996

Imrie, Alex, Pictorial History of the German Army Air Service, London 1971

Marsa-Wona Taschenatlas vom Westlichen Kriegsschauplatz, Königswartha, o.A.

Nachrichtenblatt der Luftstreitkräfte, Februar 1918

Paasche's Spezialkarten der Westfront, Blatt 3, 4, 6, 11 und 12, Verlag Paasche & Lutz, Stuttgart o.A.

Potempa, Harald, Die königlich-bayerische Fliegertruppe 1914 -1918, Frankfurt 1997

Revell, Alex, High in the Empty Blue. The History of 56. Squadron, RFC/RAF 1916 to 1920, Flying Machine Press 1995

Schwake, Norbert, Deutsche Soldatengräber in Israel, 2008

Stegemann, Hermann, Geschichte des Krieges, Band 4, Stuttgart/Berlin 1921

Internet

http://pilotesdechasse.over-blog.com/article-coadou-marcel-100684007.html

Archival records

Bavarian Main State Archive, Department 4, Munich

Angermund, Walter, KrStL 18144/29; KrStL 18016/9

Artmann, Ludwig, KrStR 18018/155; KrStR 17943/424

Bärenfänger, Karolus, OP 6166; KrStL 18016/6

Bauer, Josef, KrStR 18018/174

Böhning, Hans, OP 5021; KrStL 18016/2

Böning, Walter OP 38021; KrStL 18016/1

Brunner, Johann, KrStR 18081/703; KrStR 18125/379

Curtius, Walter, OP 5914; KrStL18144/210; KrStL 18016/16

Danhuber, Franz-Xaver, OP 6006

Diem, Josef, KrStR 18018/14

Dumler, Karl, OP 3319; KrStL 18016/18

Emmerich, Richard, OP 33450; KrStL 18016/8

Ertl, August, OP 336

Geyer, Julius, KrStR 18018/175

Grosse, Eberhard, KrStL 18016/19

Hainzl, Rupert, KrStR 18018/176; KrStR 17943/425

Hepperlin, Erwin, KrStR 18018/177; KrStR

18210/19
Hertrich, Hans, OP 41881, KrStL 18016/7
Hopf, Karl, KrStR 17994/190; KrStR 18026/152; KrStR 18018/121
Hüttner, Xaver, KrStR 18022/8; KrStR 18098/994
Kolb, Wilhelm, OP 42943; KrStL 18016/3
Koller, Karl, KrStR 18003/422; KrStR 18098/706; KrStR 18018/152
Markert, Georg, KrStR 18018/160; KrStR 17963/3426
Martin, Franz, OP 8204; KrStL 181016/14
Neu, Karl Edmund, OP 9084, KrStL 18079/13; KrStL 18016/12
Obermeier, Hermann, OP 27232, KrStL 18079/40; KrStL 18016/11
Reinhard, Josef, KrStR 18018/170
Rostock, Amandus, OP 47359; KrStL 18016/20
Schick, Max, OP 12743; KrStL 18016/13; KrStL 18016/13
Schmid, Ludwig, OP 48908, KrStL 18016/17; KrStL18031/108
Stemmler, Gottfried, KrStR 18018/153; KrStR 18209/279; KrStR 18976/49
Veith, Maximilian, KrStL 18016/15
Vollert, Clemens, OP 17694; KrStL 18016/4
Wagner, Paul, KrStR 17987/16; KrStR 18018/147
Walther Siegfried, KrStR 18157/2554
Weig, Moritz, OP 17976, KrStL 18016/10
Wenig, Erwin, OP 29523; KrStL 18016/5
Kriegsstammliste der Jagdstaffel 76
Kriegsstammrolle der Jagdstaffel 76
Abbreviation
OP = **O**ffizier-**P**ersonalakte (Officer personnel file)
KrStL = **Kr**iegs**st**amm-**L**iste
KrStR = **Kr**iegs**st**amm-**R**olle

Rainer Absmeier Collection
Miscellaneous photos of Jagdstaffel 76

Winfried Bock Collection
Excerpt from the war diary of Jagdstaffel 19, by Erich Tornuß, ca. 1930 revised and supplemented by Dr Gustav Bock and Winfried Bock
Excerpt from the war diary of Jagdstaffel 34, by Erich Tornuß, ca. 1930 revised and supplemented by Dr. Gustav Bock and Winfried Bock
Excerpt from the war diary of Jagdstaffel 76, by Erich Tornuß, ca. 1930 revised and supplemented by Dr. Gustav Bock and Winfried Bock
Excerpt from the war diary of Jagdgeschwader I
Combat report of 15. Sqn. RAF, summarised in the daily
Intelligence Summary of the RAF ("War Diary")
Combat report of 56. Sqn. RAF, ibid.
Intelligence Summary RAF, Report Capt. J. D. Belgrave 60. Sqn. R.A.F.
Intelligence Summary RAF, Report 65. Sqn. R.A.F, ibid.
Jagdstaffel 41, summary of the aerial victories by Dr. Gustav Bock
Résumé des opérations aériennes, compilation from the French official reports by Dr. Gustav Bock.

Alex Imrie Collection
von Hippel, Hans-Joachim, handwritten notes of the conversations.

Reinhard Kastner Collection
Altmann, Ludwig, letter dated June 2, 1918
Miscellaneous photos of Jagdstaffel 76
Markert Georg, letter to his wife dated May 30, 1918
Compilation of the personnel data of the Bavarian airmen in World War I, unpublished manuscript.

Jörn Leckscheid Collection
Miscellaneous photos of Jagdstaffel 76

Bruno Schmäling Collection
Arntzen, Heinrich, Tape recordings and handwritten notes of the conversations, 1978
Bauer, Josef, Jagdstaffel 76, photos and documents
Böning, Walter, Jagdstaffel 76
Letters from 1914 until 1919.
Handwritten notes of the conversations, 1977
Information, notes and sketches on the aircraft painting details
Photo album und handwritten notes
Official Confirmation of the air victories
Böhning, Hans, Jagdstaffel 76,
Photo album und handwritten notes
Unpublished text about the story of his life
Handwritten notes on the conversations with Mrs Anges Schwörer, widow of Hans Böhning.
von Borries, Fritz, Jagdstaffel 22, photos, documents, handwritten notes, personal diary, flightlog.
Fuchs, Otto, Jagdstaffel 77, handwritten notes of the conversations 1977 - 1978
Geyer, Julius, Jagdstaffel 76, photos, documents, letters.
Kolb, Wilhelm, Jagdstaffel 76, photos, documents
Koller, Karl, letter from English captivity, 7 July 1918
Leptin, Hermann, Jagdstaffel 63, photos, documents, handwritten notes.
Excerpt from the war diary of Jagdstaffel 41, by Erich Tornuß, ca. 1930
Speidel Max, Oberst a. D., Jagdstaffel 80.
Handwritten notes of the conversations and the correspondence 1979 – 1980.
Weiß, Hans, Jagdstaffel 41, photo, documents,

handwritten notes.
Wenig, Erwin, Jagdstaffel 76, Staffelführer Jagdstaffel 80, photo, documents, handwritten notes.
Vollert, Clemens, Jagdstaffel 76, photos, documents, handwritten notes.

Greg VanWyngarden Collection

Miscellaneous photos Jagdstaffel 76
G-Report File No. G/3Brigade7, Archive G.

Tobias Weber Collection

Miscellaneous photos Jagdstaffel 76

Reinhard Zankl Collection

Miscellaneous photos Jagdstaffel 76

Made in the USA
Las Vegas, NV
26 June 2026

49877020R00124